Kouamé Juslain Romaric Kouadio

Endangered mangroves

Kouamé Juslain Romaric Kouadio

Endangered mangroves

ScienciaScripts

Cover image: www.ingimage.com

This book is a translation from the original published under ISBN 978-620-6-72897-9.

Publisher:
Sciencia Scripts
is a trademark of
Dodo Books Indian Ocean Ltd. and OmniScriptum S.R.L publishing group

120 High Road, East Finchley, London, N2 9ED, United Kingdom
Str. Armeneasca 28/1, office 1, Chisinau MD-2012, Republic of Moldova, Europe
Managing Directors: Ieva Konstantinova, Victoria Ursu
info@omniscriptum.com

Printed at: see last page
ISBN: 978-620-8-37081-7

DEDICATION

I dedicate this book to

the memory of my father, the late KOUADIO Kouassi Kouma and

to my mother, the late GBELIA Likagnéné Aimée Pauline,
may God Almighty welcome them into his paradise

To my brothers and sisters

Foreword and thanks

Initially passionate about law, with the ambition of becoming a magistrate, I discovered my vocation for geography while watching a documentary on natural phenomena broadcast by National Geographic. This experience aroused in me a deep admiration for the Earth sciences, leading me to choose geography rather than law after my baccalauréat. At university, I turned to climatology, a branch of physical geography, attracted by the study of climatic phenomena. As my studies progressed, my interest grew stronger, particularly in view of the challenges posed by climate change, which is severely affecting the environment and water resources, especially in Africa and Côte d'Ivoire. Thus, my interest in the degradation of mangroves, hence the title of my book *"Mangroves menacées" (Mangroves under threat).*

I would like to express my sincere thanks to Editions Universitaires Européennes, my supervisors and all those who have supported me throughout this adventure. My hope is that this work will serve to better understand and preserve the mangroves of Côte d'Ivoire.

Table of contents

Introduction

Mangroves are vital coastal ecosystems, recognized for their unique biodiversity and the many services they provide to the environment. However, they are increasingly threatened by global climate change, which is altering their distribution and functioning. On a global scale, rising land and ocean temperatures, attributed to the increase in greenhouse gases, have favored the occurrence of extreme weather phenomena, such as more frequent storms and intense rainfall. These events increase the risk of flooding and erosion, directly threatening mangroves. At the same time, changes in precipitation patterns and ocean acidification are also influencing marine ecosystems, notably by disrupting natural plant regeneration processes and increasing the sensitivity of mangroves to disease and invasive species. In this context of rapid change, it is essential to study the complex interactions between these climatic modifications and mangroves in order to develop appropriate conservation strategies. Scientific research into the dynamics of mangroves in the face of climate change is therefore essential to anticipate future risks and define effective protection measures (Alongi, 2008, p.115).

In Côte d'Ivoire, mangroves extend over some 80,000 hectares along the Atlantic coast, representing around 3% of West Africa's mangroves (Kouadio et al., 2013). Here, these ecosystems are particularly exposed to human pressures such as deforestation for charcoal production, land conversion for agriculture and aquaculture, and pollution from industrial and domestic activities. Climate change, such as rising sea levels and increased rainfall, is also influencing their dynamics. The Tadio lagoon, located in the south of Côte d'Ivoire, is one of the country's major mangrove sites, but it is under heavy anthropogenic pressure, notably from fishing, agriculture and logging. In addition, it is subject to hydroclimatic variations such as tides, floods and periods of drought, which directly affect the evolution of mangroves. This thesis therefore examines the hydroclimatic and human contexts influencing mangrove dynamics in Tadio lagoon, using an integrated approach combining remote sensing, field surveys, modelling and socio-economic surveys.

Chapter 1: Mangroves of Laguna Tadio and their Ecological Role

Introduction

The mangroves of the Tadio lagoon represent a unique and essential ecosystem for the region, ensuring both coastal protection against erosion and soil stability. In fact, this environment, rich in biodiversity, serves as a habitat for a remarkable diversity of flora and fauna, contributing to the local ecological dynamic. In addition, local communities benefit directly, whether through fishing, timber harvesting or the exploitation of non-timber forest products. This chapter describes the characteristics of Tadio's mangroves, their ecological, economic and social importance, and the threats they face, such as deforestation, land conversion, pollution and climate change. Recommendations for sustainable management are also put forward, given the crucial importance of mangroves in regulating climate and combating coastal erosion. In a context of accelerating environmental dynamics, particularly in West Africa and southern Côte d'Ivoire, this ecosystem needs to be studied for its resilience in the face of growing pressures. Conservation strategies are therefore needed to maintain the ecosystem services it provides and limit the ecological and socio-economic impacts of its degradation.

I- Biodiversity of the mangroves of the Tadio lagoon

1. What is a Mangrove?

1- 1-1- Presentation of the study area

Our study area is located between 5°10'32" North and 5°15'14" West in southern Côte d'Ivoire, in the Grands-Ponts and Lôh Djiboi regions. The Grand-Lahou lagoon complex comprises the Nyouzomou, Tagba, Maké and Tadjo lagoons (Figure 1). It covers an area of around 230 km2. It is connected to the ocean by a pass located at the mouth of the Bandama River, a few kilometers east of Grand-Lahou (Djadji et al., 2013; Wango et al., 2013). In this part of the Ivorian coastline, mangroves extend from the Azagny canal to Ebonou, where one of the largest mangrove stands in Côte d'Ivoire can be found, with trees that can reach over 20 m in height in places s (Djadji et al., 2013; Wango et al., 2013). The Tadio lagoon is one of the watercourses that feed and drain the entire Grands-Ponts region. In the vicinity of the Tadjo lagoon, there are several villages and watercourses. These villages are inhabited mainly by fishing and farming communities, who rely heavily on the lagoon's natural resources for their livelihood. The climate is sub-equatorial and always humid, known locally as the "Attiean climate". The Tadio lagoon is a critical ecosystem, home to a wide variety of plants, animals and fish. The area is also subject to anthropogenic pressures, such as pollution, habitat degradation and excessive exploitation of resources. Like all coastal zones, the Tadio lagoon is often exposed to risks associated with climate change, such as sea-level rise, coastal erosion and storms. A study could be carried out to assess the lagoon's vulnerability to climate change and propose appropriate adaptation measures. By studying the mangroves of the Tadio lagoon, we can better understand their ecological functioning and the mechanisms that enable them to withstand harsh environmental conditions, such as sea-level fluctuations, high salinity and sediment accumulation. This knowledge is essential for the sustainable management of this fragile ecosystem, particularly with a view to its economic exploitation. The mangroves of the Tadio lagoon are also a source of income for local populations, who use mangrove wood as a building material and for charcoal production. Studying their economic potential and sustainability is therefore crucial to ensuring responsible exploitation of these natural resources and securing the livelihoods of the communities living around the lagoon. In addition, the mangroves of the Tadio lagoon play a role in regulating the local and global climate.

Map 1: Tadio Lagoon (southern Côte d'Ivoire)

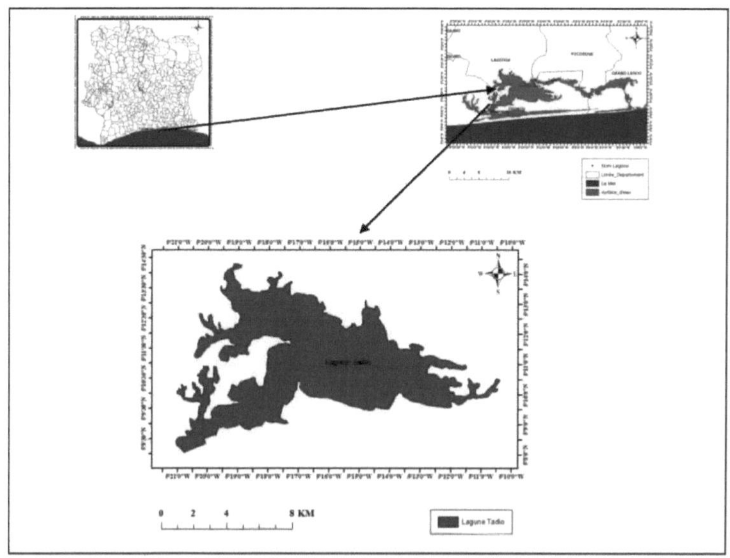

Source: Base Shp CI, Map Côte 1-2 / Production: Kouadio Romaric

1- 1-2- Presentation of mangroves: definition and general characteristics.

The mangrove, this unique and precious ecosystem, is defined from a variety of angles, depending on the field of expertise. Indeed, it is studied by ecologists, geographers and environmentalists alike, each bringing their own particular perspective to bear. This diversity of approaches makes it possible to grasp the complex essence of mangroves and appreciate their multiple functions in terms of biodiversity conservation, coastal protection and the socio-economic services they provide. Firstly, from an ecological point of view, the mangrove consists of vegetation adapted to tropical and subtropical coastal environments, located in tidal rocking zones where fresh and salt water meet. The trees and shrubs that make it up - such as mangroves of the Rhizophora, Avicennia and Sonneratia genera - are exceptionally well adapted to salinity, muddy soils and low oxygen levels. In this sense, Tomlinson (1986) describes the mangrove as a transitional ecosystem between land and sea, characterized by a root network that stabilizes the soil and promotes the growth of a specific biodiversity. Secondly, in terms of its environmental functions, mangroves stand out for their essential role in protecting coastal zones. It acts not only as a natural bulwark against erosion, storms and storm surges, but also as an ecological filter. As such, it captures sediments and pollutants, helping to purify water and maintain the quality of surrounding aquatic ecosystems. Moreover, Alongi (2009) insists that mangroves constitute a veritable "ecosystem of services", capable of supporting a diversity of marine and terrestrial species and ensuring significant carbon storage. As such, mangroves play a significant role in the fight against global warming, thanks to their ability to sequester large quantities of carbon. The biogeographical perspective also helps us to understand the global distribution of mangroves. Located mainly between latitudes 25° North and 25° South, they cover the coasts of regions such as Central and South America, West Africa, Southeast Asia and Oceania. This geographical location confers on mangroves a wide diversity of plant and animal species adapted to these singular conditions. With this in mind, Duke (1992) describes the mangrove as a cosmopolitan tropical ecosystem, highlighting the morphological and physiological adaptations that enable it to thrive despite the unstable conditions of tropical coastlines. Mangroves are also of considerable importance to local populations, who derive significant socio-economic benefits from them. It is a valuable resource in terms of timber, charcoal, medicinal products and building materials. It is also a vital fishing ground for many coastal communities. Barbier et al (2011) point out that this ecosystem, by providing fishery resources and promoting the development of aquaculture, contributes to the food and economic security of the regions that depend on it. This shows that mangroves are much more than just coastal forests: they embody a multi-faceted local heritage. Finally, from a legal and conservation standpoint, mangroves in many countries benefit from a regulatory framework designed to ensure their protection. The fragility of this ecosystem, combined with its ecological value, has led to the adoption of specific laws to prevent its degradation and promote its restoration. The FAO (2007), in its recommendations, presents the mangrove as a natural heritage to be preserved, requiring sustainable management policies and conservation strategies involving local communities. This participatory approach is deemed essential to guarantee the sustainability of mangrove ecosystems in the face of increasing human-induced pressures.

1- 2- Environmental conditions in Tadio Lagoon and mangrove reproduction processes

1- 2-1- Environmental conditions of the Tadio lagoon

The Tadio lagoon, located in southern Côte d'Ivoire, is characterized by distinct and varied environmental conditions that directly influence mangrove ecosystems. Firstly, the region's climate is typically tropical, with alternating wet and dry seasons. During the rainy season, abundant rainfall feeds the lagoon, increasing water levels and modifying salinity, which favors the growth of certain plant species adapted to these variations. Secondly, the high average temperatures, generally fluctuating between 24 and 30 degrees Celsius, create an environment conducive to the development of mangroves. This constant warmth, combined with high humidity, contributes to efficient plant photosynthesis and the proliferation of plant biomass. Tides also play a crucial role in the dynamics of the Tadio lagoon. The regular ebb and flow of the tides replenishes the water, preventing stagnation and ensuring the continuous circulation of nutrients essential to the local flora and fauna. This tidal movement also helps to disperse mangrove seeds, facilitating the natural regeneration of the mangroves. In addition, the lagoon's soil, composed mainly of fine sediments and organic matter, provides a fertile substrate for mangroves to root in. The nutrient-rich composition of the soil promotes rapid plant growth, while the water-holding capacity of the sediments ensures constant water availability, even during periods of low tide. In addition, water quality in Tadio lagoon is influenced by various anthropogenic and natural factors. Freshwater inflows from surrounding rivers partially dilute the salinity, creating ideal conditions for a variety of mangrove species that cannot tolerate excessively high salinity levels. Nevertheless, human activities such as agriculture and fishing can introduce additional pollutants and sediments, affecting overall water quality and ecosystem health. In conclusion, the environmental conditions of Tadio lagoon are defined by a complex interplay between tropical climate, tides, soil composition and water quality. These factors interact to create a unique habitat that supports rich biodiversity and provides vital ecosystem services.

1- 2-2-Process and mode of reproduction of mangroves in the Tadio lagoon

The mode of reproduction of mangroves in the Tadio lagoon is a complex and fascinating phenomenon, which deserves to be studied in depth. First of all, it should be remembered that mangroves are dioecious plants, meaning that there are male and female plants. Pollination is therefore an essential step in the reproductive process, and is generally carried out by wind or insects. Then, once pollination has taken place, the mangrove seeds, or propagules, develop on the mother plant. These propagules are often elongated and bean-shaped, and are endowed with special characteristics that enable them to survive and thrive in a marshy, saline environment. For example, the propagules of some mangrove species are able to float and germinate while still in the water, a considerable advantage in a constantly changing environment. Then, when the propagules are mature, they are released by the mother plant and begin their journey through sea and river currents. This dispersal stage is crucial to the survival and diversity of mangroves, enabling them to colonize new environments and maintain genetic diversity. Propagules can travel long distances, which is another important factor in dispersal and colonization of new habitats. Finally, propagules that have reached a favorable environment, such as the Tadio lagoon, will settle and develop, forming a new generation of mangroves. Young mangrove seedlings will seek out loose substrates, such as silt, which will enable them to develop their roots optimally. These roots, which are often aerial, have several functions: they enable the plant to maintain itself in an unstable environment, absorb oxygen from the air and promote the exchange of nutrients. The way in which the mangroves of the Tadio lagoon reproduce is a true testament to nature's resilience and ingenuity, a complex and dynamic process that enables these plants to survive and thrive in a particularly demanding environment. It's also important to note that the way mangroves reproduce is closely linked to their life cycle. Mangroves are perennial plants, which means they can live for many years, even decades. Their life cycle is therefore characterized by a growth and development phase, followed by a reproduction and dispersal phase. The growth and development phase is particularly important for mangroves, as it enables them to adapt to their environment and develop the characteristics they need to survive and reproduce. For example, mangroves are able to tolerate significant variations in salinity, which is a considerable advantage in an environment like the Tadio lagoon. They are also able to develop aerial roots, which enable them to anchor themselves in loose substrates and absorb oxygen from the air. The reproduction and dispersal phase of mangroves is also crucial to their survival and diversity. Reproduction and dispersal of propagules enable them to colonize new habitats and maintain a certain genetic diversity. This genetic diversity is important, as it enables mangroves to adapt to new environments and cope with threats such as disease or climate change. The way mangroves reproduce in the Tadio lagoon is a complex and fascinating phenomenon, closely linked to their life cycle and adaptation to their environment.

1- 3- Mangrove formation in Tadio lagoon

1- 3-1-Geological factors in the formation of mangroves in Tadio lagoon

Geological factors play a central role in the formation and development of the mangroves of the Tadio lagoon, a unique and precious ecosystem located on the Ivorian coast. To better understand the influence of these factors, it is essential to analyze various aspects such as soil type, topography, geomorphological processes and regional tectonic movements. Firstly, soil type is a key factor in the establishment and growth of mangroves. These particular forests require muddy soils rich in organic matter to thrive. In the case of the Tadio lagoon, the sediments brought in by rivers and tides provide an ideal substrate for the development of these ecosystems. Indeed, sediment deposits, made up of fine particles and decomposing organic matter, provide an environment conducive to the rooting and growth of mangrove trees, the emblematic trees of mangroves. Secondly, the topography of the site plays a decisive role in the formation of mangroves. The shallow, flat coastal zones characteristic of the Tadio lagoon are ideal for mangrove installation and growth. These topographical conditions also allow good water circulation, essential to the survival and expansion of these forests. In addition, tide-induced water level variations help to create specific micro-habitats, offering a diversity of spaces for the colonization of different mangrove species. In addition, geomorphological processes such as erosion and sedimentation actively shape and form the mangroves of Tadio lagoon. The combined action of tides, waves and ocean currents deposits sediment and creates new areas suitable for mangrove colonization. This sedimentation phenomenon enables the gradual accumulation of material, thus promoting the expansion of mangroves into new areas. In addition, the erosion of banks and islets can free up space for mangrove expansion, while providing additional sediment for the growth and development of these ecosystems. In addition, regional geological factors, such as plate tectonics and vertical movements of the coastline, exert an influencc on the formation and evolution of mangroves in the Tadio lagoon. These phenomena can influence coastal subsidence or emergence, thereby modifying environmental conditions and favoring or limiting mangrove development. For example, a subsidence movement can lead to an increase in the floodable surface, offering new opportunities for mangrove colonization. Conversely, an emergence movement may reduce the area available for mangroves, limiting their expansion. Finally, it is important to consider the interactions between geological factors and other environmental factors, such as climate, salinity and biodiversity, which jointly influence the formation and development of mangroves in Tadio Lagoon. The balance between these different factors contributes to the creation of a complex and dynamic ecosystem, capable of adapting to change and maintaining its resilience in the face of disturbance.

1- 3-2- Pedological characteristics in the formation of mangroves in the Tadio lagoon

Soil characteristics play a fundamental role in the formation and development of the mangroves of the Tadio lagoon, a remarkable ecosystem located on the Ivorian coast. To gain a deeper understanding of the influence of these characteristics, it is essential to examine various aspects such as soil texture, structure, chemical composition and biological properties. Firstly, soil texture is a key element in the establishment and growth of mangroves. These unique forests require muddy, fine-particle soils to thrive. In the case of the Tadio lagoon, the sediments brought in by rivers and tides provide an ideal substrate for the development of these ecosystems. The silt and clay deposits, mixed with decomposing organic matter, provide an ideal environment for mangrove trees to take root and grow. Secondly, soil structure plays a decisive role in mangrove formation. Well-structured soils, with their stable aggregates and interconnected pores, encourage the circulation of water and air, essential to the survival and expansion of these forests. In addition, tide-induced water level variations help to create specific soil conditions, offering a diversity of spaces for colonization by different mangrove species. In addition, the chemical composition of soils has a major influence on the formation and development of mangroves in Tadio lagoon. Mangrove soils are generally acidic to neutral, with a PH between 4 and 8. Salinity is also an important factor, as mangroves are able to tolerate high levels of salt in the soil. This salt tolerance is due to physiological and morphological adaptations in mangroves, enabling them to excrete or store excess salt. In addition, the presence of nutrients such as nitrogen, phosphorus and potassium is essential for mangrove growth and development. The processes of mineralization and decomposition of organic matter contribute to the release of these nutrients into the soil, making them available to plants. In addition, the biological properties of mangrove soils in the Tadio lagoon play a crucial role in the formation and evolution of these ecosystems. Organic matter, derived from the decomposition of leaves, roots and other plant residues, is an important source of nutrients for mangroves. It also helps to improve soil structure and water retention, thus promoting mangrove growth and expansion. Soil microbial biodiversity, including bacteria, fungi and other microorganisms, also plays an essential role in nutrient recycling and pollutant degradation. Finally, it is important to consider the interactions between soil characteristics and other environmental factors, such as climate, geology and biodiversity, which jointly influence the formation and development of mangroves in the Tadio lagoon. The balance between these different factors contributes to the creation of a complex and dynamic ecosystem, capable of adapting to change and maintaining its resilience in the face of disturbance.

1- 3-3-Ecological factors in the formation of mangroves in Tadio lagoon

Ecological factors play a key role in the formation and development of the mangroves of the Tadio Lagoon, a highly valued ecosystem on the Ivorian coast. To better understand the influence of these factors, it is necessary to analyze various aspects such as climate, plant and animal species, biotic interactions and ecological processes. First of all, climate is a key factor in the establishment and growth of mangroves. These unique forests develop in tropical and subtropical regions, where temperatures are high and rainfall abundant throughout the year. In the case of the Tadio lagoon, the warm, humid climate offers ideal conditions for the development of these ecosystems. Indeed, high temperatures favor the growth and reproduction of mangrove trees, the emblematic trees of the mangroves, while rainfall provides the freshwater necessary for their survival. Secondly, the plant and animal species present in the mangroves of the Tadio lagoon actively contribute to the formation and evolution of these ecosystems. Mangroves, adapted to the particular living conditions of intertidal zones, are the main component of these forests. Their aerial roots, known as pneumatophores, absorb oxygen and ensure the stability of muddy soils. Mangroves are also home to a wide variety of animal species, such as crustaceans, molluscs, fish and birds, which interact with mangroves and contribute to the overall functioning of the ecosystem. Biotic interactions also play a key role in the formation and development of mangroves in the Tadio lagoon. Symbiotic relationships, such as those established between mangroves and nitrogen-fixing bacteria, contribute to soil nutrient enrichment and promote plant growth. In addition, trophic relationships, such as predation and parasitism, regulate animal populations and maintain ecosystem equilibrium. Competitive interactions between plant species, notably for access to light, water and nutrients, also influence the structure and composition of mangroves. In addition, ecological processes such as organic matter decomposition and nutrient dynamics actively contribute to the formation and evolution of mangroves in Tadio Lagoon. The decomposition of plant litter and animal residues by soil microorganisms releases nutrients essential for mangrove growth and development. Tides also play a crucial role in nutrient dynamics, ensuring transport and redistribution within the ecosystem. Finally, it's worth noting the interactions between ecological factors and other environmental factors, such as geological and pedological factors, which jointly influence the formation and development of mangroves in the Tadio lagoon. The balance between these different factors contributes to the creation of a complex and dynamic ecosystem, capable of adapting to change and maintaining its resilience in the face of disturbance.

2- Plant species present

2- 1- Shrub species

The mangroves of the Tadio lagoon are made up of different types of trees and plants adapted to salty, muddy coastal environments. During our surveys, we were able to observe a few, such as Rhizophora, which is characterized differently. Indeed, trees of this genus, such as Rhizophora mangle (red mangrove) and Rhizophora racemosa (white mangrove), are the dominant species in the mangroves of Tadio lagoon. They have aerial roots called "stilts" that enable them to rise above the water. Rhizophora species are the most common in Tadio lagoon. However, they enable the species to survive in environments regularly flooded by the tides. The stilts of the aerial roots rise above the tidal level, enabling the plant to obtain oxygen directly from the air. This is crucial, as the swamp soils in which mangroves grow are often oxygen-poor. This adaptation also enables the plant to withstand frequent tidal flooding. The stilts of the aerial roots enable the plant to remain above water, reducing the risk of root asphyxiation. In addition to these aerial roots, the plant also possesses other adaptations in its evolution. For example, it can filter salt from seawater through glands on its leaves, enabling it to thrive in salty environments. The plant is also able to reproduce by producing propagules that can float on water and take root when they reach a suitable environment. Next we have the Avicennia species, notably Avicennia germinans (black mangrove), also found in the mangroves of the Tadio lagoon. They have slightly differentiated aerial roots called "pneumatophores" that help supply oxygen to the submerged roots. In addition to these species, there's also the Laguncularia racemosa, also known as the white mangrove, a salt-tolerant tree found in the mangroves of the Tadio lagoon. It is distinguished by its heart-shaped leaves and less-developed aerial roots. Avicennia lenticels are small pore-like structures on the roots, stem and sometimes even the leaves. These pores enable the roots to take in oxygen from the ambient air, even when partially submerged. This adaptation is crucial for these plants, as they often thrive in muddy, water-saturated soils. Avicennia's thick, rough leaves are another important adaptation. These leaves reduce water loss due to evaporation by limiting the surface area exposed to the air and creating a thicker cuticular barrier. This enables the plant to conserve water and resist water stress. These physiological and morphological adaptations enable Avicennia to survive in harsh environments where other plants cannot thrive. These evolutionary features are an example of the adaptability of species to changing environmental conditions, and their importance in the survival of mangrove ecosystems. White mangrove (Laguncularia racemosa) is a widespread mangrove species in the coastal regions of the Tadio lagoon in Côte d'Ivoire. This shrubby or arborescent plant can grow up to 8 meters high and is distinguished by its oval, leathery leaves, which are dark green on top and white and velvety on the underside. White mangroves are adapted to life in the extreme conditions of mangroves, where they must cope with significant variations in salinity, humidity and water level. They have aerial roots called pneumatophores, which enable them to breathe in water-saturated soils. These trees play an essential role in the mangrove ecosystems of the Tadio lagoon in Côte d'Ivoire. They help stabilize coastal soils, protect coastlines against erosion and storms, filter water and provide habitats for numerous plant and animal species. In addition, white mangroves are invaluable to local communities who depend on the lagoon's natural resources for their livelihoods. They provide firewood, building materials, fruit and traditional medicines. However, the white mangroves of the Tadio lagoon are also threatened by deforestation, pollution and coastal development. It is therefore crucial to implement conservation and sustainable management measures to preserve these unique ecosystems and their biodiversity. Conocarpus erectus, also known as buttonwood, is a shrub or small tree found in the mangroves of the Tadio lagoon. It can tolerate drier conditions than other mangroves and is often found on the edges of mangroves. The buttonwood has developed physiological and morphological adaptations that enable it to survive in extreme environments, such as the wet and saline zones of mangroves.

Physiologically, buttonwood is able to tolerate high concentrations of salt in the soil and surrounding water. It has developed desalination mechanisms, notably by excreting salt through its leaves, retaining water to dilute salt, and filtering salt through its roots. Morphologically, buttonwood has features that help it to withstand the harsh conditions of mangroves. For example, its leaves are thick, leathery and often covered with hairs, which reduces water loss through evaporation and protects cells from salt damage. In addition, its aerial roots, called pneumatophores, enable it to adapt to the low oxygen content of mangrove soil. These roots develop from the base of the tree and rise above the water level, creating a network of air channels that allow the tree to breathe. Buttonwood plays an important role in mangrove ecosystems. Its roots help stabilize the soil, preventing coastal erosion. It also provides shelter and a food source for many animals, such as birds, fish and crabs. The mangroves of the Tadio lagoon are also home to the Bruguiera, which is regularly found on the Mangrove de rivage. Bruguiera's aerial branches develop roots that hang down to the muddy mangrove soil. These aerial roots, known as stilts, help the plant to rise above the salty, muddy water, enabling its leaves to receive sufficient light for photosynthesis. The aerial roots also provide structural support for the plant in unstable mangrove conditions. Also, Bruguiera seeds still germinate on the tree before falling into the water. This adaptation enables young plants to root immediately in the mud, increasing their chances of survival. Viviparous seeds are also adapted to water dispersal, as they can float to areas suitable for mangrove growth. Finally, there's the oyster. The mangrove oyster of the Tadio lagoon is characterized by its tolerance of variable salinity, enabling it to survive in waters with fluctuating conditions. Its robust, irregular shell offers protection against predators and makes it easy to anchor to mangrove roots. It also plays an essential role in water filtration, improving the quality of the surrounding ecosystem. It also contributes to biodiversity by being a source of food for numerous species. Finally, the oyster is of economic importance to the local communities that harvest it.

2- 2- Ferns in the mangroves of Tadio lagoon

The mangroves of the Tadio lagoon are a unique ecosystem, rich in biodiversity. Among the many plant species growing there are ferns, which have adapted to this particular environment. First of all, ferns are vascular plants characterized by their absence of flowers and seeds. They reproduce by means of spores, which are contained in structures called sporangia. Ferns are also known for their ability to thrive in difficult conditions, such as nutrient-poor soils and shady areas. In the mangroves of the Tadio lagoon, several species of fern can be found. One of the most common is the sword fern, or Nephrolepis biserrata. This fern has sword-shaped fronds that can reach up to a metre in length. It often grows in higher areas of the mangrove, where it can benefit from a little more sunlight. Another species of fern that can be found in the mangroves of the Tadio lagoon is the Boston fern, or Nephrolepis exaltata. This fern has shorter, wider fronds than the sword fern, and often grows in the shadier areas of the mangrove. It is also known for its ability to purify the air by absorbing pollutants. In addition to these two species, rarer ferns can also be found in the mangroves of the Tadio lagoon. For example, the crested fern, or Doryopteris concolor, is a species characterized by its crested fronds. It often grows in the wettest areas of the mangrove. Ferns play an important role in the mangrove ecosystem. They help maintain soil moisture and prevent erosion. They also provide a habitat and food source for numerous animal species, such as insects and birds. However, ferns are also threatened by human activities. Destruction of mangrove habitat, water pollution and overexploitation of natural resources are all factors that can affect the survival of ferns in the mangroves of the Tadio lagoon. The mangroves of the Tadio lagoon are home to a wide variety of ferns that have adapted to this particular environment.

Photo 1: Mangrove species in Tadio Lagoon

2- 3-Herbaceous species present in the mangroves of the Tadio lagoon

The herbaceous species present in the Tadio lagoon, located in the south of Côte d'Ivoire, are diverse and each has distinct characteristics that contribute to the richness of this ecosystem. In fact, these species are diversified by their characteristics. Firstly, the spartina grass (Spartina alterniflora) stands out for its ability to tolerate high levels of salinity and flooding. It has robust underground rhizomes that help stabilize soil and reduce erosion. This species often forms dense colonies, creating important habitats for local wildlife, including birds and small invertebrates. Secondly, paspalum (Paspalum vaginatum) is renowned for its tolerance of salinity and its ability to adapt to variations in water level. Its leaves are long and thin, and it often grows in dense clumps. This grass plays a crucial role in stabilizing banks and filtering nutrients, thus contributing to the quality of the lagoon's water. Papyrus cyperus (Cyperus papyrus), or papyrus, is an emblematic wetland species. It is characterized by its triangular stems and umbellate inflorescences. Papyrus forms dense mats that provide shelter and food for many aquatic species. In addition, its roots help to filter and purify water, improving the quality of the aquatic environment. Eleocharys dulcis (Eleocharis dulcis), also known as yellow nutsedge, is notable for its cylindrical stems and edible tubers. This plant is well adapted to flooded environments and helps stabilize wet soils. The tubers are also a source of food for local wildlife and human communities. Finally, cane grass (Panicum repens) is characterized by its creeping rhizomes and long, narrow leaves. This grass is particularly resistant to salinity and water stress. It plays an important role in preventing soil erosion and providing habitat for insects and small animals. However, these herbaceous species of the Tadio lagoon are crucial to maintaining the ecological balance of this wetland. Each makes unique contributions to the stability, water filtration and biodiversity of this coastal ecosystem.

3- Species zonation

3- 1- High tide zone

The high tide zone in the mangroves of the Tadio lagoon is an area that is regularly submerged by high tide waters. This zone is subject to particular environmental conditions, such as high levels of salinity, strong exposure to waves and currents, and low oxygen availability in the soil. The mangrove species that grow in the high tide zone are adapted to these harsh environmental conditions. Red mangroves (Rhizophora mangle) are the dominant species in this zone, due to their ability to tolerate high levels of salinity and strong exposure to waves and currents. These trees have stilt roots and pneumatophore roots, which enable them to anchor themselves in the soil and capture oxygen from the air. Red mangroves in the high tide zone are generally smaller in size and stockier in shape than those growing in areas further out to sea. This is due to the harsher environmental conditions in this zone, which limit tree growth and height. The leaves of red mangroves in this zone are also thicker and more resistant than those of trees growing in areas further from the sea, enabling them to withstand strong winds and salt spray. In addition to red mangroves, other mangrove species can also be found in the high tide zone, although they are generally less numerous and less dominant. Black mangroves (Avicennia germinans) and white mangroves (Laguncularia racemosa) are species that can tolerate slightly lower salinity levels than red mangroves, and may therefore be present in the high tide zone. The high tide zone is an important area in the mangroves of Tadio lagoon, as it plays a key role in protecting the coast from erosion and flooding. The stilt and pneumatophore roots of the red mangroves in this zone help stabilize the soil and reduce the impact of waves and currents. In addition, the mangroves in this area provide an important habitat for many animal species, such as crabs, fish and birds.

The high tide zone in the mangroves of Tadio lagoon is an area subject to harsh environmental conditions, such as high levels of salinity and strong exposure to waves and currents. The dominant species in this area are red mangroves, which are adapted to these conditions and play a key role in protecting the coast and providing habitat for animal species. Other mangrove species may also be present in this zone, although they are generally less numerous and less dominant.

3- 2-Intermediate zone

The intermediate zone in the mangroves of the Tadio lagoon is an area located between the high and low tide zones. This zone is subject to variable environmental conditions, as it is regularly submerged by the waters of high tide, but is also exposed to air and sun during low tide. The mangrove species that grow in the intermediate zone are adapted to these changing environmental conditions. Red mangroves (Rhizophora mangle) are always present in this zone, but they are usually accompanied by other mangrove species, such as black mangroves (Avicennia germinans) and white mangroves (Laguncularia racemosa). These species have different tolerances to salinity and flooding, enabling them to coexist in the intermediate zone.

Black mangroves are particularly suited to the intermediate zone, as they can tolerate high levels of salinity and flooding. They have pneumatophore roots that enable them to capture oxygen from the air, and thick, tough leaves that enable them to withstand strong winds and salt spray. White mangroves are more tolerant of flooding than red mangroves, but less tolerant of salinity. They have stilt roots that enable them to anchor themselves in the soil and capture oxygen from the air. In the intermediate zone, tree height is generally higher than in the high tide zone, due to the more favorable environmental conditions for tree growth. However, tree height can vary considerably depending on species and location. Black mangroves tend to be the tallest, while white mangroves are generally smaller. Trees near the high tide zone tend to be smaller and stockier than those near the low tide zone. In addition to mangrove species, the intermediate zone is also home to a wide varicty of animal species, such as crabs, fish and birds. The roots and trunks of the trees provide an important habitat for these species, as well as a source of food. The intermediate zone in the mangroves of the Tadio lagoon is a zone subject to changing environmental conditions, which is reflected in the structure and species composition of the mangroves. Red, black and white mangroves are all present in this zone, with different tolerances to salinity and flooding. Tree heights vary according to species and location, and the intermediate zone is also home to a wide variety of animal species.

3- 3- Low tide zone

The low-tide zone in the mangroves of the Tadio lagoon is an area that is regularly exposed to air and sunlight during low tide, but is also submerged by the waters of high tide. This area is subject to variable environmental conditions, such as changing salinity levels and shorter flooding periods than in the high tide zone. The mangrove species that grow in the low-tide zone are adapted to these changing environmental conditions. Red mangroves (Rhizophora mangle) are always present in this zone, but they are usually accompanied by other mangrove species, such as white mangroves (Laguncularia racemosa) and round-leaved mangroves (Conocarpus erectus). These species have different tolerances to salinity and flooding, enabling them to coexist in the low-tide zone. White mangroves are particularly well adapted to the low-tide zone, as they can tolerate shorter periods of inundation than other mangrove species. They have stilt roots that enable them to anchor themselves in the soil and capture oxygen from the air. Round-leaf mangroves are also adapted to the low-tide zone, as they can tolerate higher salinity levels than white mangroves. They have thick, tough leaves that enable them to withstand strong winds and salt spray. In the low-tide zone, tree height is generally higher than in the high-tide zone, due to the more favorable environmental conditions for tree growth. However, tree height can vary considerably depending on species and location. White mangroves tend to be the tallest, while round-leaved mangroves are generally smaller. Trees near the intermediate zone tend to be taller and stockier than those near the high tide zone. In addition to mangrove species, the low-tide zone is also home to a wide variety of animal species, such as crabs, fish and birds. The roots and trunks of the trees provide an important habitat for these species, as well as a source of food. The low-tide zone in the mangroves of Tadio lagoon is an area subject to changing environmental conditions, which is reflected in the structure and composition of mangrove species. Red, white and round-leaf mangroves are all present in this zone, with different tolerances to salinity and flooding. Tree heights vary according to species and location, and the low-tide zone is also home to a wide variety of animal species.

II- Evolution of mangrove plant masses in Tadio lagoon

1- Fauna associated with mangroves and their adaptation to the environment

1- Fauna associated with mangroves

1- 1- Avifauna and aquatic fauna

The avifauna associated with the mangroves of the Tadio lagoon is rich and diverse, with many bird species using this habitat for feeding, breeding and resting. Mangroves offer a variety of bird habitats, including woodland, marsh and intertidal areas. The birds associated with the mangroves of Tadio Lagoon are often adapted to life in wetland and coastal areas. For example, many bird species have developed long legs and beaks to feed in shallow waters and mudflats. Mangroves are also home to many threatened and endangered bird species, including herons, spoonbills and ibises. The most common bird species in the mangroves of the Tadio lagoon include the grey heron (Ardea cinerea), the spoonbill (Platalea leucorodia), the African kingfisher (Halcyon senegalensis), the black-headed gull (Larus genei), the Caspian tern (Hydroprogne caspia) and the interrupted-necked plover (Charadrius alexandrinus). Mangroves are also an important resting place for migratory birds, such as the redshank (Tringa totanus) and black-tailed godwit (Limosa limosa).

The aquatic fauna associated with the mangroves of the Tadio lagoon is also rich and diverse, with many species of fish, crustaceans and molluscs using this habitat for feeding, reproduction and protection. Mangroves offer a variety of habitats for aquatic fauna, including marsh areas, intertidal zones and submerged mangrove areas. Fish associated with the mangroves of the Tadio lagoon are often adapted to life in the shallow, brackish waters of the mangroves. For example, many fish species have developed enlarged pectoral fins to move around in shallow water, as well as specialized organs to regulate the salt content in their bodies. The most common fish species in the mangroves of Tadio Lagoon include tilapia (Oreochromis spp.), mullet (Mugil spp.), barracuda (Sphyraena spp.), catfish (Clarias spp.) and bighead carp (Hypophthalmichthys nobilis). Secondly, crustaceans associated with the mangroves of Tadio Lagoon include species such as shrimps, crabs and lobsters. Mangroves provide an important habitat for crustaceans, particularly as a breeding and feeding ground. The most common crustacean species in the mangroves of Tadio lagoon include grey shrimp (Penaeus spp.), long-antenna shrimp (Macrobrachium spp.), fiddler crab (Uca spp.) and mud crab (Scylla serrata). Finally, molluscs associated with the mangroves of the Tadio lagoon include species such as oysters, mussels and sea snails. Molluscs play an important role in the mangrove ecosystem as filter feeders and decomposers. The most common mollusc species in the mangroves of the Tadio lagoon include the mangrove oyster (Crassostrea spp.), the mangrove mussel (Brachidontes spp.) and the sea snail (Littorina spp.). The avifauna and aquatic fauna associated with the mangroves of the Tadio lagoon are rich and diverse, with many species using this habitat for feeding, reproduction and protection. Mangroves offer a variety of habitats for wildlife, including wooded areas, marshland and intertidal zones.

1- 2- Insects, Reptiles and Amphibians

The mangroves of the Tadio lagoon, located on the west coast of Côte d'Ivoire, are a unique and diverse ecosystem, home to a wide variety of animal species. These include insects, reptiles and amphibians, which play an important role in maintaining the ecological balance of the mangrove. Insects are one of the most diverse and important groups of fauna associated with the mangroves of the Tadio lagoon. They include species such as mosquitoes, flies, butterflies, dragonflies and ants. Mosquitoes and flies are often considered nuisances, but they play an important role in the mangrove ecosystem as pollinators and food sources for other animals. Butterflies and dragonflies are important predators of small insects, helping to regulate their populations. Ants are important decomposers, helping to break down leaf litter and organic matter into plant nutrients. Reptiles are another important group of fauna associated with the mangroves of Tadio Lagoon. The most common species include snakes, lizards and turtles. Snakes are important predators of small mammals, birds and fish in the mangroves. Lizards are also important predators of small insects, helping to regulate their population. Turtles are important herbivores, feeding on mangrove leaves and stems, helping to maintain the structure and health of the mangrove. Amphibians are a less diverse group in the mangroves of Tadio Lagoon, but they nevertheless play an important role in the ecosystem. The most common species include frogs and toads. Frogs are important predators of small insects, helping to regulate their population. Toads are also important predators of small insects and worms, helping to maintain the health of the mangrove soil. The insects, reptiles and amphibians associated with the mangroves of Tadio Lagoon are often adapted to life in wetlands and coastal areas. For example, many insect species have developed longer wings and legs to move around in shallow water and mudflats. Reptiles such as snakes and lizards have developed scales and membranes to protect themselves from salt water and predators. Amphibians such as frogs and toads have developed specialized skins and lungs to survive in wet and flooded areas. However, the fauna associated with the mangroves of the Tadio lagoon is threatened by numerous human activities, including deforestation, pollution and climate change. Deforestation of mangroves for agriculture, urbanization and aquaculture leads to habitat loss and fragmentation of the animal population. Pollution of mangrove water and soil by industrial and domestic waste, fertilizers and pesticides, can lead to animal death and disease. Climate change, particularly rising sea levels and more frequent and intense storms, can lead to habitat loss and disruption of the mangrove ecosystem. The insects, reptiles and amphibians associated with the mangroves of the Tadio lagoon are an important and diverse group of animal fauna, playing an important role in maintaining the ecological balance of the mangrove.

2- Ecological relationships, competition and cooperation between fauna and mangroves

2- 1-Ecological relationships between wildlife and mangroves

Mangroves are unique ecosystems vital to the health of our planet. Located in the tidal zones of tropical and subtropical regions, these salty swamp forests are home to a wide diversity of plant and animal species, while providing numerous ecological, economic and social benefits. In Côte d'Ivoire, the Tadio lagoon is one of the most important sites for mangroves, with an area of over 4,000 hectares. In this chapter, we will explore the ecological relationships between wildlife and mangroves in the Tadio lagoon, focusing on the different animal groups and their interaction with the mangrove ecosystem. The mangroves of the Tadio lagoon are made up of several tree and shrub species adapted to harsh environmental conditions, such as saline soils, regular flooding and strong winds. The foliage of the mangrove species in the Tadio lagoon is therefore specially adapted to cope with these conditions and maximize their survival and growth. The different mangrove species have unique foliage characteristics that play an important role in the development of the associated fauna. Invertebrates are an important group of animals associated with the mangroves of Tadio Lagoon. The most common species include crabs, shrimps, molluscs and worms. These animals play an important role in the mangrove ecosystem as decomposers, predators and filter feeders. Crabs are one of the most important and visible groups of invertebrates in the mangroves of the Tadio lagoon. They are often referred to as the "engineers of the ecosystem" because of their role in modifying the mangrove habitat. Crabs dig burrows in the mangrove soil, helping to aerate and circulate water in the soil. They also feed on dead leaves and organic matter, contributing to the decomposition and recycling of nutrients in the ecosystem. Fish are another important group of animals associated with the mangroves of the Tadio lagoon. Mangroves provide an essential habitat for many fish species, notably as a breeding and feeding ground. Mangroves also provide a refuge for fish due to their complex root and trunk structure, which offers protection from predators and extreme environmental conditions. Fish associated with the mangroves of Tadio Lagoon are often adapted to life in the shallow, brackish waters of the mangroves. For example, many species of fish have developed enlarged pectoral fins to move around in shallow water, as well as specialized organs to regulate the salt content in their bodies. Birds are another important group of animals associated with the mangroves of Tadio Lagoon. Mangroves offer a variety of habitats for birds, including woodland, marsh and intertidal areas. Mangroves are also located along migratory flyways, making them an important resting place for migratory birds. The birds associated with the mangroves of Tadio Lagoon are often adapted to life in wetland and coastal areas. For example, many bird species have developed long legs and beaks to feed in shallow waters and mudflats. Mangroves are also home to many threatened and endangered bird species, including herons, spoonbills and ibises. Reptiles and amphibians are less diverse groups of animals in the mangroves of Tadio Lagoon, but they nevertheless play an important role in the ecosystem. The most common species include snakes, lizards, turtles and frogs. Reptiles and amphibians are often adapted to life in wetlands and coastal areas, with specialized skins and lungs to survive in flooded areas.

Reptiles and amphibians are important predators of small insects, worms and crustaceans in mangroves, helping to regulate their populations. Finally, mammals are a less common group of animals in the mangroves of Tadio Lagoon, but they are present nonetheless. The most common species include monkeys, rats and bats. Mammals are often adapted to life in woodland and swamp areas, with specialized legs and tails for moving through flooded areas. Mammals are important predators of small animals in the mangrove, helping to regulate their population. The ecological relationships between wildlife and mangroves in Tadio lagoon are complex and interdependent. Different groups of animals have important roles to play in the mangrove ecosystem, as decomposers, predators, filter feeders and ecosystem engineers. Mangroves offer a variety of animal habitats, including woodland, marsh and intertidal zones. Animals associated with the mangroves of Tadio Lagoon are often adapted to life in wetland and coastal areas, with unique physical and behavioral characteristics to maximize survival and growth.

2- 2- Competition and cooperation between wildlife and mangroves

Mangroves are complex and dynamic ecosystems, where the relationships between flora and fauna are many and varied. In the Tadio lagoon, mangroves are home to a wide variety of animal species, which interact in different ways with the plants around them. Indeed, competition is an ecological relationship in which two or more species compete for the same limited resource. In the mangrove swamps of the Tadio lagoon, competition for space and nutrients is common between different species of plants and animals. For example, crabs and shrimps can compete for space and food resources in the holes and crevices of mangrove roots. Fish and molluscs can also compete for space and nutrients in marsh and intertidal zones. However, competition is not always harmful to the species involved. In some cases, competition can lead to species specialization and diversification, which can increase the biodiversity and biological productivity of the ecosystem. For example, the different mangrove species in the mangroves of the Tadio lagoon have different growth and reproduction strategies, enabling them to coexist and specialize in different ecological niches. Furthermore, cooperation is an ecological relationship in which two or more species interact in a mutually beneficial way. In the mangroves of the Tadio lagoon, cooperation between fauna and plants is commonplace and essential to the maintenance of the ecosystem. For example, crabs and shrimps play an important role in the decomposition and recycling of nutrients in mangroves, feeding on dead leaves and organic matter. Crab and shrimp droppings are also rich in nutrients, which can benefit the surrounding plants. Birds and bats are also important partners for the mangroves of the Tadio lagoon. Birds and bats feed on the fruits and seeds of the various mangrove species, which contributes to seed dispersal and mangrove regeneration. Bird and bat droppings are also rich in nutrients, which can benefit the surrounding plants. Finally, the cooperative relationships between wildlife and mangroves in the Tadio lagoon may also involve more complex and subtle interactions. For example, mangrove roots provide a habitat and food source for many species of bacteria and fungi, which in turn can help plants absorb nutrients and resist disease. Animals such as earthworms and termites can also play an important role in the decomposition and recycling of nutrients in mangroves, interacting with bacteria and fungi in the soil. In short, the competitive and cooperative relationships between fauna and mangroves in Tadio lagoon are complex and interdependent. Different animal groups have important roles to play in the mangrove ecosystem, as decomposers, predators, filter feeders, pollinators and seed dispersers. Competitive relationships can lead to species specialization and diversification, which can increase the biodiversity and biological productivity of the ecosystem. Cooperative relationships are essential for the maintenance of the mangrove ecosystem, particularly for nutrient decomposition and recycling, seed dispersal and disease resistance.

3- Dynamics of mangroves in the Tadio lagoon (southern Côte d'Ivoire)

3- 1- Mangrove vegetation index in Tadio lagoon

NDVI, an index commonly used to quantify vegetation in a specific area, is calculated from satellite images by comparing the reflectance of the near-infrared and red spectra. This index varies between -1 and 1: high values indicate dense, healthy vegetation, while values close to 0 indicate bare soil or water, and negative values may indicate artificial surfaces.

- ✓ **1990 vegetation index**

The Normalized Difference Vegetation Index (NDVI) map for 1990, with values fluctuating between -0.296296 and 0.52459, reveals valuable information about the state of vegetation cover at that time. Firstly, observation of the red zones, which represent the lowest NDVI values, highlights very low or even absent vegetation cover. These negative or near-zero values may be associated with bare surfaces, degraded soils or areas devoid of dense vegetation. In addition, these areas could correspond to bodies of water, impervious surfaces or zones subject to marked deforestation. Next, green areas, corresponding to high NDVI values, reflect denser, healthier vegetation. With values as high as 0.52459, these areas indicate the presence of significant vegetation cover, such as forests, wetlands rich in biodiversity, or well-managed farmland. As a result, these areas demonstrate notable ecological vitality, suggesting favorable environmental conditions, more sustainable land management or less human disturbance. On the other hand, it is crucial to note that intermediate zones, shown in orange or yellow, indicate moderate NDVI values, which could reflect sparse or regenerating vegetation. These intermediate variations are often characteristic of savannahs, grasslands or lands in ecological transition. The 1990 NDVI map shows a significant disparity in vegetation distribution. Low-index areas highlight environmental challenges such as deforestation or soil degradation, while high-index areas reveal pockets of dense, healthy vegetation. This map is therefore an essential tool for understanding the initial state of the vegetation, and for guiding conservation and restoration strategies in the years to come.

Map 2: NDVI (Normalized Difference Vegetation Index) 1990

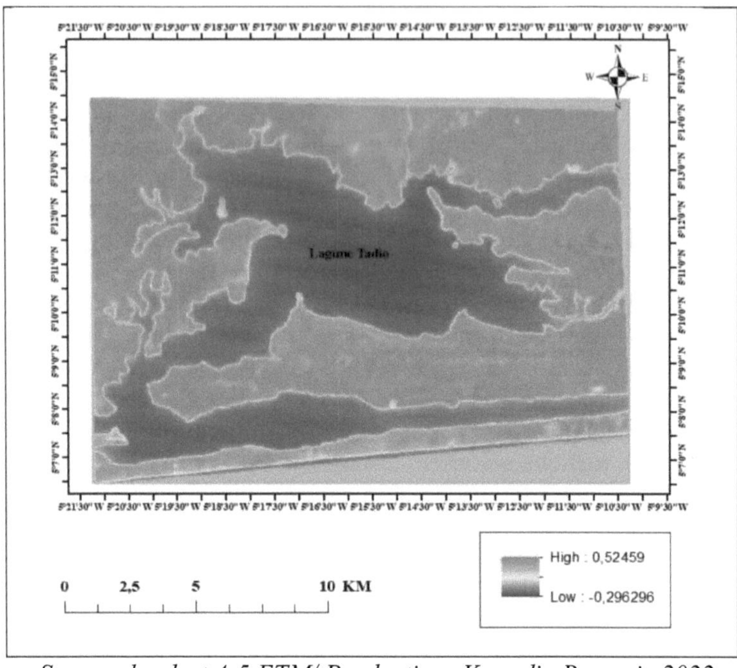

Source: landsat 4-5 ETM/ Production: Kouadio Romaric 2022

✓ **The 2000 vegetation index**

The NDVI map for the year 2000, with values ranging from -0.0592605 to 0.0281647, illustrates a worrying vegetation situation in the region studied. From the outset, it should be emphasized that the values of this index are extremely low, revealing a very low vegetation density throughout the territory represented. This low amplitude of values betrays an ecosystem in difficulty, marked by a notable deterioration in vegetation cover. First of all, negative values, close to -0.0592605, appear in the red areas. This indicates the presence of surfaces devoid of dense vegetation, such as bare soil, degraded terrain, or areas where vegetation has been completely eliminated. It is also possible that these areas include water bodies or artificial spaces, where natural vegetation has been largely replaced or eliminated. Next, the positive values, which reach a maximum of just 0.0281647, are represented by shades of light green. This reflects the presence of extremely sparse vegetation of low vigour. These values, well below the thresholds observed for healthy vegetation, suggest unfavorable ecological conditions, such as poor soils, inadequate land management or negative climatic impacts that have led to low biological productivity. Furthermore, it is important to note that intermediate zones, usually indicative of transitions between different ecosystems, are virtually absent here. This reflects a landscape uniformly affected by widespread degradation, with no significant transition zones to denser or more vigorous vegetation. In other words, the map shows a worrying homogeneity, characterized by a very low density of vegetation cover.

Map 3: NDVI (Normalized Difference Vegetation Index) 2000

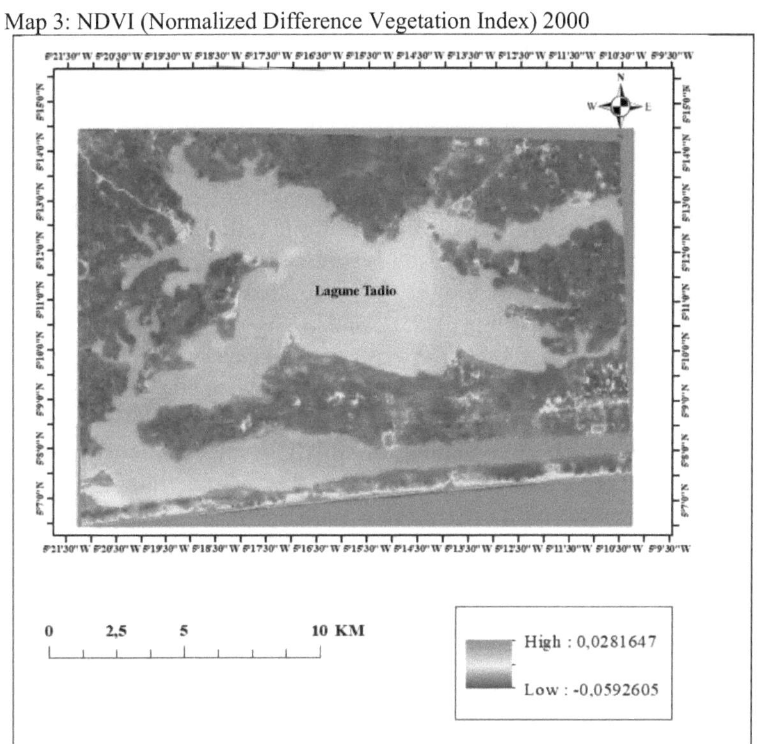

Source: Landsat 7 ETM / Production: Kouadio Romaric 2022

✓ The 2022 vegetation index

The NDVI map for the year 2022, showing an index variation between -0.124986 and 0.144589, provides an overview of recent vegetation dynamics in the study area. Firstly, it is important to note that index values remain relatively low, although higher than those observed in 2000. This suggests a slight improvement in vegetation cover, but also that ecosystems have not yet returned to flourishing health. Firstly, negative index values, fluctuating around -0.124986, appear in red. These values indicate areas where vegetation is either absent or severely degraded. These areas could correspond to bare soil, severely eroded land, or artificial surfaces such as human infrastructures. Furthermore, the persistence of these negative values reflects the inability of these areas to regenerate, underlining the need for ecological rehabilitation measures. Positive values, up to 0.144589, are represented by shades of green. This improvement on 2000 shows that some areas have experienced vegetation regeneration, albeit still modest. These areas of denser vegetation could correspond to secondary forests, restored wetlands or agricultural areas where sustainable management practices have been implemented. It is also possible that these improvements are due to human interventions such as reforestation or the implementation of conservation measures. On the other hand, intermediate values, ranging from yellow to orange, indicate areas in transition where vegetation is present but still weak. These areas could represent land undergoing restoration, where natural regeneration is progressing slowly, or semi-natural landscapes where human impact is still perceptible. Thus, these values reflect an ecosystem in the process of recovery, but one that remains vulnerable to environmental disturbance. In short, analysis of this 2022 NDVI map reveals a slightly improved environmental situation compared to 2000, although challenges remain. The continued presence of sparsely vegetated areas underlines the region's ecological fragility, while the modest progress observed in some parts indicates a possible resilience if adequate measures are maintained or reinforced. All in all, this map is a valuable tool for assessing the restoration efforts undertaken and for guiding future sustainable land management strategies.

Map 4: NDVI (Normalized Difference Vegetation Index) 2022

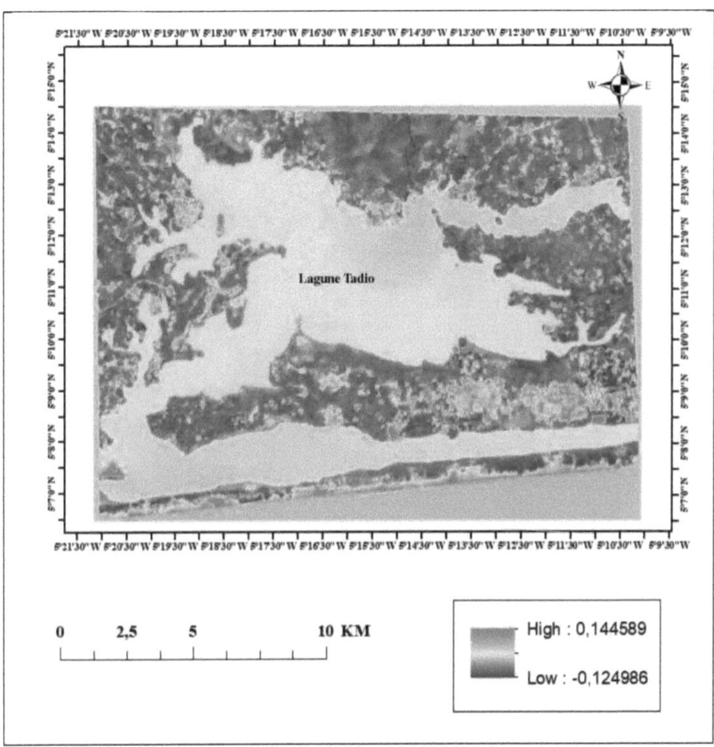

Source: Landsat 8 OLI / Production: Kouadio Romaric 2022

3- 2- Mangrove dynamics in Tadio Lagoon from 1990 to 2022

A comparative analysis of land use between 1990, 2000 and 2022 in the Tadio lagoon region highlights significant dynamics that testify to the environmental and anthropogenic transformations of this area.

✓ Vegetation dynamics in the vicinity of Tadio Lagoon in 1990

For 1990 vegetation, we can see that dense forests have the highest proportion of land. This suggests that the study area has relatively high forest cover during this period, which may be beneficial for biodiversity and ecosystem conservation. Next, we have degraded forests, as well as crops and fallow land, which account for a small proportion of the area during this year. This reflects the low level of human activity in the area at this time, when people have not yet begun to show more interest in activities such as agriculture and other environmental degradation projects. On the other hand, bare soil or dwellings account for a large proportion of the area, indicating a certain human pressure on the environment. Finally, we have observed an evolutionary growth in mangroves compared with other years.

Map 5: Vegetation dynamics of the Tadio lagoon perimeter 1990

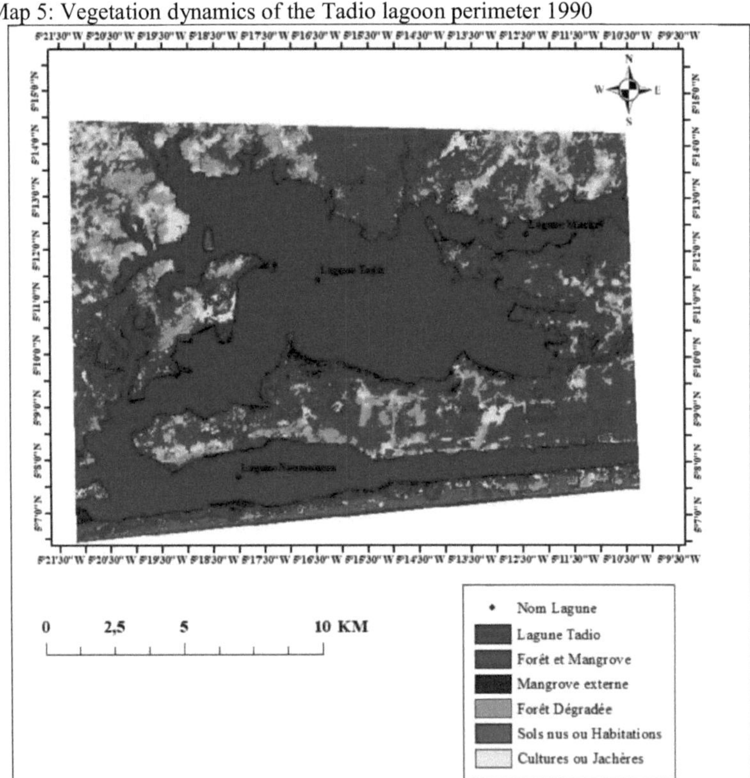

Source: landsat 4-5 ETM / Production: Kouadio Romaric 2022

- ✓ **Vegetation dynamics in the vicinity of Tadio Lagoon in 1990**

The year 2000 saw a slightly accelerated deterioration in vegetation. During this year, we observed a dense forest that was beginning to degrade, giving way to a degraded forest, a large area of cultivated or fallow land and a large area of bare soil or housing. It should be noted that this degradation of the vegetation cover is reflected in the high level of occupation of the area by the population. So, unlike in previous years, during this period, people are beginning to devote themselves a little more to agriculture. As for the dynamics of the mangroves during this year, we note an average evolution compared to the previous year. Although the vegetation is degrading, giving way to a degraded forest, the dynamics of the mangroves are a little more remarkable, suggesting several reasons for their evolution.

Map 6: Vegetation dynamics of the Tadio 2000 lagoon perimeter

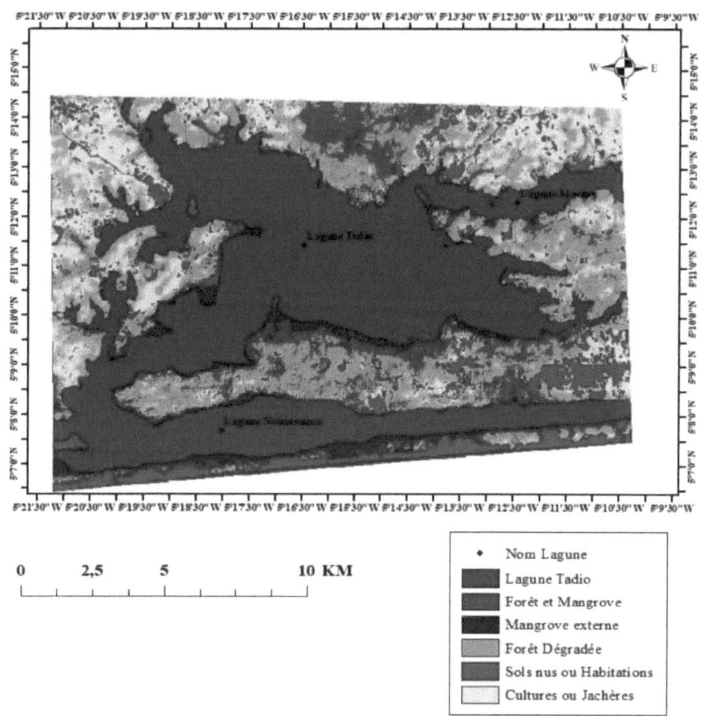

Source: Landsat 7 ETM/ Production: Kouadio Romaric 2022

- ✓ **Vegetation dynamics in the vicinity of Tadio Lagoon in 1990**

The 2022 image analyzed shows an average vegetation cover dynamic for the Tadio lagoon perimeter. In recent years, we have observed an accelerated change in the vegetation cover of our study area. We have a forest with a small surface area. Naturally, this low surface area translates into a high surface area of degraded forest, as well as crop and fallow land. In addition to these large areas, there is a remarkable amount of bare soil or housing in the zone. It should also be noted that a period of fallow land can be the result of natural soil regeneration, crop rotation or the restoration of biodiversity. These processes help to maintain soil fertility, promote plant growth and restore the balance of local ecosystems. This situation explains the transformation of bare soil into cultivated or fallow land. However, we see a strong dynamic of mangroves during this period, which can be explained by their adaptation to specific environmental conditions, their role as natural protection against degradation and their positive effects on local biodiversity.

Map 7: Vegetation dynamics of the Tadio 2022 lagoon perimeter

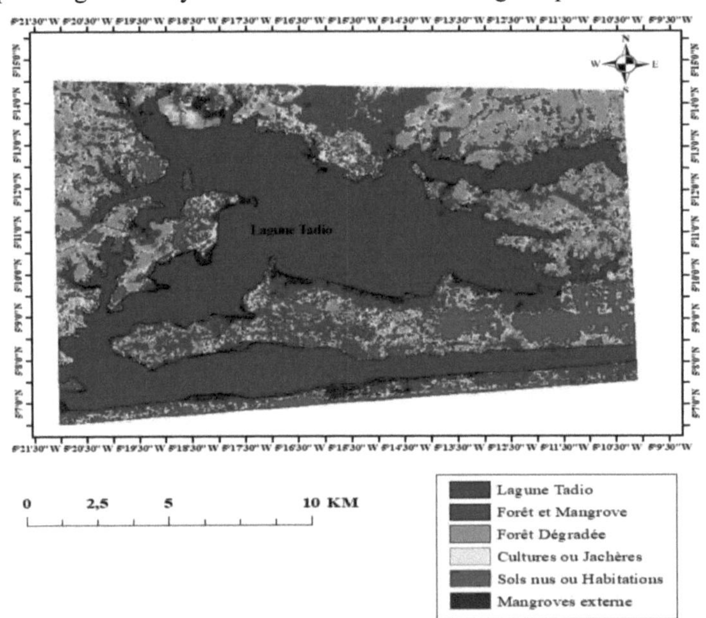

Source: Landsat 8 Oli/ Production: Kouadio Romaric 2022

3-3- Statistics and proportion of land use and external mangrove dynamics from 1990 to 2022

The statistical table of land cover between 1990 and 2022 highlights significant changes in the distribution of the different land cover categories. Firstly, in 1990, the "Water" category occupied 40% of the total surface area, followed by "Mangrove and forest" at 39%, or 107 km^2 . These two categories dominate, accounting for 79% of the surface area. However, other categories such as "External mangrove" (3%), "Bare soil or dwellings" (5%) and "Crops or fallow land" (5%) occupy much smaller areas. Degraded forest", at 8%, already indicates significant anthropic pressure.

In 2000, the surface area of the "Water" category increased to 42%, while "External Mangrove" rose from 3% to 5%. However, "Mangrove and forest" underwent a sharp reduction, reaching only 13%, a worrying change probably linked to deforestation and urbanization. Conversely, "Bare soil or dwellings" and "Crops or fallow land" increase to 10% and 12% respectively, illustrating increasing urbanization and agricultural expansion. Degraded forest" almost doubles its share to 18%, a worrying indicator of increasing forest degradation.

Finally, in 2022, the proportion of the "Water" category remains stable at 41%, while "External Mangrove" declines slightly to 4%. Mangrove and forest", on the other hand, shows a notable recovery, expanding to 22%, which can be attributed to reforestation or conservation initiatives. Bare soil or housing" continues to rise, to 13%, underlining sustained demographic pressure. However, "Crops or fallow land" is down to 8%, suggesting a shift in agricultural activities. Finally, "Degraded forest" is down to 12%, perhaps indicating restoration efforts, although the situation remains worrying for the sustainable management of natural resources.

In summary, analysis of the proportions between 1990 and 2022 reveals complex dynamics in which certain natural categories, such as water bodies and forests, remain dominant despite increasing anthropogenic pressures, while urbanization and land degradation continue to transform the local landscape.

Table 1: Land use statistics from 1990 to 2022

	1990		*2000*		*2022*	
Surface	***Percentage***	***Area KM2***	***Percentage***	***Area KM2***	***Percentage***	***Area KM2***
Water	*40%*	*109*	*42%*	*114*	*41%*	*113*
Outer mangrove	*3%*	*7*	*5%*	*13*	*4%*	*8*
Mangrove and forest	*39%*	*107*	*13%*	*36*	*22%*	*60*
Bare floors or homes	*5%*	*13*	*10%*	*28*	*13%*	*37*
Crops or fallow land	*5%*	*14*	*12%*	*34*	*8%*	*22*
Degraded forest	*8%*	*22*	*18%*	*50*	*12%*	*34*
Total	*100%*	*273*	*100%*	*273*	*100%*	*273*

III- Importance of mangroves in the context of climate change

1- Mangroves as carbon sinks and reducing coastal erosion

1- 1- Mangroves as carbon sinks

The carbon cycle in the mangroves of the Tadio lagoon is a complex ecological process, integrating the exchange, storage and transformation of carbon between the atmosphere, vegetation, soil and water. As productive coastal ecosystems, mangroves play an active role in carbon sequestration and storage, contributing to the global carbon cycle. Firstly, photosynthesis enables plants such as mangroves to absorb carbon dioxide from the atmosphere and transform it into organic matter. This organic matter is then used for plant growth and cellular respiration, which releases CO_2 back into the atmosphere. In addition, a portion of this organic matter is transmitted to other elements of the ecosystem, notably through leaves and branches falling to the ground, where their decomposition releases CO_2 while enriching the soil with humus. Roots also play a role in the carbon cycle by releasing dissolved organic compounds into the water, feeding aquatic micro-organisms and sustaining the lagoon's biodiversity.

What particularly distinguishes the carbon cycle in the mangroves of the Tadio lagoon is their ability to sequester carbon over the long term. Indeed, mangrove soils, rich in organic matter, retain this carbon for millennia thanks to the anoxic conditions fostered by regular flooding. However, various disturbances, both natural and man-made, compromise this storage capacity. Natural events, such as storms, can erode mangroves and release accumulated carbon. Similarly, human activities, including deforestation and pollution, alter ecological processes, reducing the resilience of this essential ecosystem.

1- 2- Role of mangroves in reducing coastal erosion

Mangroves, essential for preserving coastlines, stabilize soils and reduce erosion through a number of mechanisms. Firstly, their robust roots anchor themselves deeply in the soil, creating a dense network that ensures resistance to waves and storms. As they grow, these roots consolidate the substrate and encourage the formation of particle aggregates through the secretion of organic substances, reinforced by the action of micro-organisms. They also slow down currents, reducing their erosive power and facilitating sedimentation. In addition, the decomposing leaves and branches of mangroves form an organic layer that traps sediment and enriches the soil. By reducing the force of waves and creating an environment favorable to biodiversity, this structure plays an active role in widening the coastline. Similarly, in the Tadio lagoon, mangroves attenuate wave energy by creating a natural barrier and promote the accumulation of sediments that stabilize the banks. Ultimately, the interaction between these processes not only consolidates the coastline but also extends it, confirming the crucial importance of protecting these ecosystems to ensure coastal resilience and environmental sustainability.

1- 3- Recycling mangrove nutrients in Tadio lagoon

Because of their biodiversity and the biological processes they undergo, the mangroves of the Tadio lagoon ensure optimal recycling of nutrients, essential for their productivity. Firstly, the decomposition of organic matter is facilitated by detritivorous bacteria, fungi and invertebrates, which transform this matter into simple molecules, followed by its mineralization into inorganic elements. This release of nutrients supports plant growth and the dynamism of the ecosystem. Environmental conditions specific to mangroves, such as alternating tides and salinity, also influence the efficiency of these processes, although excessive accumulation can lead to eutrophication or acidification, which are harmful to biodiversity.

In addition, bacteria, fungi and invertebrates play a key role in the degradation of nutrients, making them available to other organisms. Bacteria, for example, transform matter into assimilable molecules, while fungi degrade more resistant matter, such as lignin. Invertebrates, by fragmenting matter, accelerate the decomposition process.

Secondly, the biological fixation of nitrogen and phosphorus by bacteria and micro-organisms interacting with plants also contributes to nutrient recycling and ecosystem productivity. This process enables plants to compensate for the low availability of these elements in the environment, thereby enhancing their growth.

Finally, nitrogen, phosphorus and other essential nutrients (potassium, calcium, magnesium, etc.) from various sources (water, sediments, human activities) ensure mangrove productivity. Transformed by geochemical and biological processes, these elements feed plants and organisms and contribute to the ecosystem's ecological functions. Thanks to their role in photosynthesis, respiration and other processes, these nutrients ensure the resilience and preservation of the biodiversity of the mangroves of the Tadio lagoon.

2- Health of mangroves in Tadio lagoon

2- 1- Seasonal health of mangroves

The mangroves of the Tadio lagoon are ecologically important ecosystems, subject to seasonal variations that influence their state of health in distinct ways depending on the season. During the rainy season, increased precipitation alters freshwater and saltwater inputs, influencing salinity and promoting the growth of certain mangrove species. Rainfall also provides essential nutrients, although in excess it can lead to the proliferation of invasive plants. In addition, rainfall conditions support the natural reproduction and regeneration of mangroves, strengthening their resilience.

During the dry season, on the other hand, reduced rainfall limits available water and nutrients, increasing salinity and posing osmotic and desiccation constraints. Some species then develop adaptive strategies, such as mycorrhizal symbiosis, to compensate for the low availability of nutrients, and conserve water and nutrients in order to maintain a certain level of activity.

The intermediate seasons, between rainy and dry periods, feature moderate precipitation that causes fluctuations in salinity and water level, influencing soil stability and root oxygenation. Nutrient availability depends on biogeochemical cycles, and species adopt adjusted biological schedules to reproduce and regenerate during this transition.

In short, each season affects mangrove growth, productivity and resilience in unique ways. It is essential to integrate these seasonal variations into strategies for the preservation and sustainable management of these ecosystems.

2- 2- Diseases and pests of mangroves in Tadio lagoon

As ecologically important coastal ecosystems, the mangroves of the Tadio lagoon are vulnerable to a variety of diseases. First and foremost, bacterial diseases, such as root rot caused by Vibrio bacteria and bacterial leaf gall caused by Pseudomonas, affect plant productivity and health. To counter these infections, integrated management based on water quality, biodiversity and the reduction of human disturbance is required.

On the other hand, viral diseases, although less frequent, cause considerable damage, for example with the yellow mosaic virus, which limits leaf growth. Their spread can be halted by monitoring vectors and appropriate biological treatments. Parasitic diseases, caused by nematodes, insects or crustaceans, weaken mangrove roots and leaves, increasing their vulnerability. Finally, non-infectious diseases, caused by environmental factors such as drought or pollution, lead to water stress and plant mortality. Management of the underlying causes and control of water and soil quality are therefore essential to limit their impact.

As for pests, crustaceans and molluscs also cause damage by feeding on plant tissue. Crabs, for example, damage leaves and roots, while certain gastropods reduce photosynthesis by destroying leaves and seeds. Faced with these nuisances, a strategy of prevention and protection of mangroves is essential for their sustainable conservation.

3- Interactions between diseases, pests and environmental stressors

3- 1-Synergistic effects of diseases, pests and environmental factors

Mangroves in the Tadio lagoon are exposed to a variety of environmental stressors, as well as diseases and pests, which can affect their health and productivity. The synergistic effects of these stressors can have serious consequences for mangrove health and the ecosystem as a whole. Environmental stressors, such as pollution, salinity, storms and flooding, can weaken mangroves and make them more vulnerable to disease and pests. For example, water and soil pollution can reduce nutrient availability and affect mangrove growth, making them more susceptible to pest attack and pathogen infection. Diseases and pests can also interact to cause greater damage to mangroves. Plants weakened by disease can be more vulnerable to pest attack, and conversely, pest damage can facilitate pathogen infection. For example, mangrove crabs can damage mangrove bark and leaves, which can facilitate infection by pathogenic fungi or bacteria. The synergistic effects of disease, pests and environmental stressors can have serious consequences for the health of mangroves in Tadio Lagoon. Damage to plants can reduce their growth, reproduction and ability to store carbon, which can affect the structure and function of the ecosystem as a whole. In addition, weakened mangroves may be less resilient to environmental disturbances, such as storms and floods, leading to biodiversity loss and habitat degradation.

3- 2- Mangrove resistance and resilience to diseases and pests

Over time, ecosystems have developed resistance and resilience mechanisms that enable them to cope with these stress factors and maintain their ecological integrity. First and foremost, the biodiversity of mangroves in the Tadio lagoon plays a crucial role in their resistance to disease and pests. Indeed, a greater diversity of mangrove species and associated organisms reduces the likelihood of massive pest infestations and the spread of disease. This is due to the presence of resistant species and natural predators that help maintain an ecological balance and limit the impact of stress factors. In addition, the mangroves of the Tadio lagoon have developed physiological defense mechanisms against diseases and pests. Some mangrove species produce chemical compounds, such as tannins and phenols, which can inhibit the growth of pathogens and discourage pests. In addition, mangroves can strengthen their plant tissues and cell walls to protect themselves against pest and disease damage. These defense mechanisms enable mangroves to resist stress factors and maintain their health and productivity. The resilience of mangroves in the Tadio lagoon to disease and pests is also linked to their ability to regenerate and recover from disturbance. Mangroves have a high reproductive and dispersal capacity, enabling them to rapidly recolonize degraded or damaged areas. In addition, mangroves can store reserves of nutrients and energy in their roots and stems, enabling them to recover rapidly from periods of stress. This capacity for regeneration and recovery is essential for maintaining the ecological integrity of mangroves and ensuring their long-term sustainability. Finally, the management and conservation of mangroves in the Tadio lagoon can strengthen their resistance and resilience to disease and pests. Management and conservation measures include restoring degraded habitats, promoting biodiversity, setting up disease and pest monitoring and control programs, and reducing environmental stressors such as pollution and overexploitation of resources. These measures aim to preserve the ecological integrity of mangroves and strengthen their capacity to cope with threats and stress factors.

3-3- The impact of diseases and pests on mangrove biodiversity and ecosystem services

Biodiversity and coastal protection are exposed to a variety of threats, including disease and pests. The consequences of these stressors on mangrove biodiversity and ecosystem services can be significant and long-lasting. Firstly, diseases and pests can affect mangrove biodiversity by reducing species richness and abundance. Indeed, diseases and pests can cause plant and animal mortality, which can lead to a reduction in species diversity and a loss of the ecological functionality of mangroves. In addition, diseases and pests can alter the structure and composition of species communities, which can have repercussions on ecological interactions and food webs. Moreover, diseases and pests can alter the ecosystem services provided by mangroves. Mangroves are productive ecosystems that provide numerous ecosystem services, such as protecting coastlines against erosion and flooding, sequestering carbon, filtering water and providing habitats for numerous species. Disease and pests can affect these ecosystem services by reducing the productivity and health of mangroves. For example, diseases and pests can reduce the capacity of mangroves to store carbon, filter water and protect coastlines from erosion and flooding. In addition, disease and pests can have economic and social impacts on local communities that depend on mangroves for their livelihoods. Mangroves provide important resources for local communities, such as firewood, timber, seafood and fish. Disease and pests can reduce the availability of these resources, impacting on the livelihoods and food security of local communities. Finally, diseases and pests can interact with other environmental stressors, such as pollution, climate change and resource overexploitation, to amplify their impacts on mangrove biodiversity and ecosystem services. For example, diseases and pests can weaken mangroves, making them more vulnerable to the impacts of climate change, such as rising sea levels and more frequent storms. The consequences of disease and pests for the biodiversity and ecosystem services of mangroves in the Tadio lagoon can be significant and long-lasting.

Conclusion

In short, this chapter highlights the importance of mangroves in the Tadio lagoon, both ecologically and socially. Firstly, the exceptional biodiversity of mangroves, illustrated by the variety of plant and animal species, underlines their crucial role in ecosystem balance. What's more, the lagoon's specific environmental conditions contribute to the reproduction and formation of mangroves, reinforcing their resilience in the face of climate change. Furthermore, the evolution of plant masses demonstrates the complex interactions between flora and fauna, as well as the impact of ecological relationships on mangrove dynamics. Analyses of the vegetation index reveal significant variations from 1990 to 2022, highlighting the importance of ongoing monitoring of these ecosystems. Mangroves also play a key role as carbon sinks and in reducing coastal erosion, offering solutions to current environmental challenges. Their state of health, influenced by seasonal factors and the presence of diseases and pests, calls for appropriate conservation measures to ensure their long-term survival. The conclusions of this chapter underline not only the biological richness of mangroves in the Tadio lagoon, but also their essential contribution to the sustainability of coastal ecosystems in the face of anthropogenic and climatic pressures. This finding calls for in-depth reflection on the need to preserve these fragile ecosystems for future generations.

Chapter 2: Impacts of Hydroclimatic Changes on Mangroves in Lagune Tadio

Introduction

This chapter examines climatic parameters to detect variations in temperature, precipitation, humidity, atmospheric pressure, wind and solar radiation, which are essential for understanding the climatic evolution of the study area. At the same time, it helps to forecast future conditions using climate models and to anticipate changes. In addition, thermal and rainfall variations in the Tadio lagoon, analyzed over various time periods, are prompting researchers to assess their impacts in order to provide valuable environmental information to the population. The study focuses on the evolution of rainfall and temperature between 1983 and 2022, including monthly and annual variations in these elements. By observing the temporal distribution of rainfall indices, it clarifies their specific features. In addition, Chapter III explores the interactions between oceanic indices and mangroves, which are sensitive to maritime variations, demonstrating the influence of tides, currents, salinity and temperature on their health, distribution and biodiversity. The use of measurement tools and methods, as well as the challenges of data collection and analysis, enrich this exploration. The study aims to illustrate the oceanic influence on sustainable mangrove management. Mangroves, essential to coastal biodiversity, depend on freshwater for their development. The Tadio lagoon clearly demonstrates the effects of this, through rainfall, rivers and groundwater, all of which guarantee the survival of these ecosystems. Freshwater nourishes and moderates salinity, influencing tides, sedimentation, erosion and species distribution. With this in mind, this chapter looks in detail at freshwater sources and their role in the resilience of Tadio Lagoon's mangroves, emphasizing their preservation to maintain their unique ecosystem services.

I- Climate Change and Evolving Hydroclimatic Conditions

1- Analysis of climatic parameters

1- 1- Analysis of monthly rainfall and temperature trends

1- 1-1-Analysis of monthly rainfall trends

The rainfall pattern in the study area is characterized by alternating wet and dry seasons, organized in a bimodal fashion: a long rainy season from April to July, followed by a short dry season in August and September, then a short rainy season in October and November, and finally a long dry season from December to March. In this analysis, the rainy months show significant rainfall between April and July, as well as in October-November, varying according to the stations observed. Rainfall peaks in July during the main rainy season, with monthly totals generally fluctuating between 110 and 380 mm. This rainfall, though intense, fluctuates from station to station.

The dry period lasts approximately six to seven months a year and is accompanied by low rainfall, particularly in December and January, with amounts ranging from 16 to 63 mm depending on the station. The harmattan phenomenon, characterized by a dry wind from the Sahara, accentuates the drought from November to February, further reducing rainfall, particularly during the transitional months. This dry wind, although normally attenuated in this southern region, is seeing its effects amplified by climate change, contributing to increased variability in rainfall patterns. However, given the geographical position of our study area, harmattan should not be felt in principle, but due to the climatic phenomenon we observe it especially during the months of December, January and February. The effects of the harmattan will not be the same as in the North. In our area, this will be reflected in a drop in rainfall during this period and a few moments marked by dry winds.

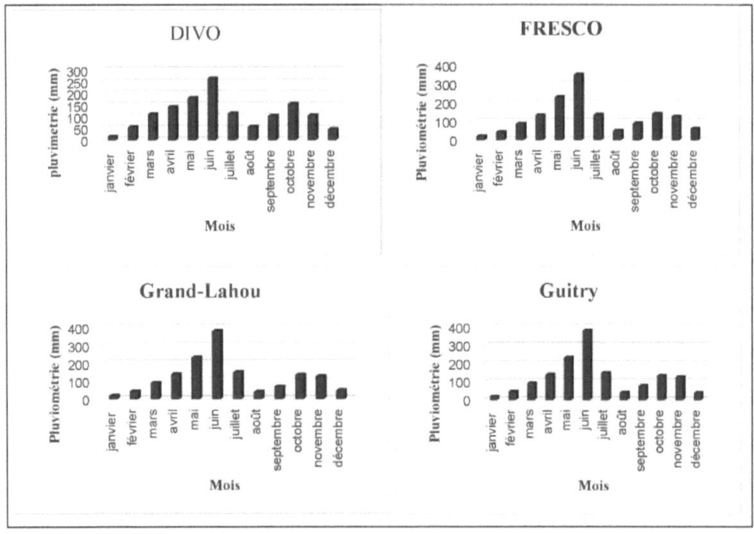

Data source: SODEXAM, NASA / Production: Kouadio Romaric 2021

Figure 1: Monthly rainfall trends

1- 1-2- Thermal average analysis

This cold period is marked by very low temperatures, varying between 25°C and 26°C for all observation stations in the study area. In fact, this period is characterized by good rainfall throughout the region. The three départements have temperature heights that are not homogeneous, and the months that characterize this cold period are May 25°C, June 25°C and July 25°C. The stations observed have practically the same temperature variations. However, the months of the short rainy season and the transition months more often record a temperature high of 27°C or 28°C. In fact, this period occurs at the end of the long rainy season and during the short rainy season. However, it sometimes happens that the transition months are influenced by the preceding dry season, making it difficult to identify this average temperature period. The months that mark this average temperature period are March, April, September, October and November.

In the bimodal regime, the dry season is also marked by a rise in temperatures in the south of the country. We have a hot period characterized by high temperatures, sometimes reaching 28°C. In other words, these temperatures vary between 27°C and 28°C. This period marks the start or presence of the dry season. There's no rainfall, and even if there is, it's orographic rainfall, and this rain, after falling on the ground, causes a kind of heat, due to the vapour after precipitation, because of the high ground temperature during this dry period. The warmest months are December, January and August. This situation of thermal heights is identical from one station to another. The transitional months that mark the beginning or end of the rainy season, such as March and November, often record the warmest temperatures during this season.

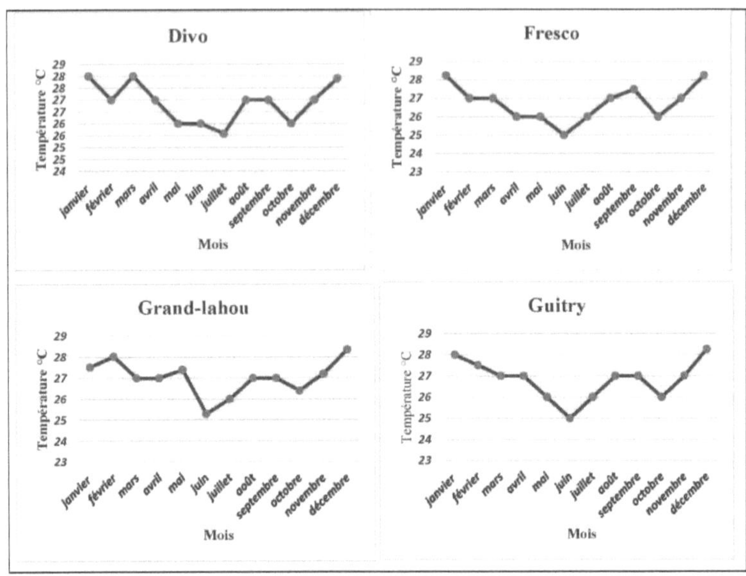

Data source: SODEXAM, NASA / Production: Kouadio Romaric 2022
Figure 2: Monthly temperature trend

1- 2- Analysis of interannual rainfall and temperature variations

1- 2-1- Interannual rainfall trends

The analysis of interannual rainfall values shows the evolution of rainfall heights at the various observation stations in our study area. Analysis of the values shows a variation in rainfall over the time series. Indeed, we observe an average evolution at the beginning of the time series analysis, with rainfall amounts in 1983 being approximately higher than in recent years. In addition, each observation station records different rainfall values, with rainfall values varying between 1100 mm and 1800 mm depending on the year. Generally speaking, the rainfall amounts analyzed are on average the same. In recent years, there has been a marked change in rainfall levels in our study area. The stations have recorded high rainfall amounts, especially over the last ten years, with rainfall amounts of over 1800 mm. However, it is important to note that the increase in rainfall in recent years is due to several factors. There are several factors that contribute to the evolution of rainfall over the years. These include climate change, which can have an impact on rainfall patterns. For example, global warming can lead to changes in atmospheric circulation, which can influence the distribution and intensity of precipitation. Other factors, such as ocean circulation patterns, variations in solar activity, local topography or extreme weather events, can play a role in the evolution of rainfall.

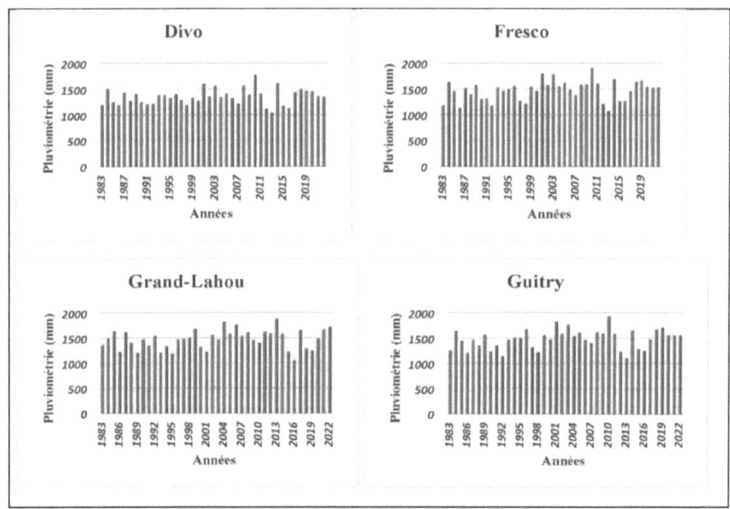

Data source: SODEXAM, NASA / Production: Kouadio Romaric 2021

Figure 3: Interannual rainfall trends

1- 2-2- Interannual temperature trends

The figures below show temperature trends in the study area. Our various stations analyzed show temperatures of up to 28°C. Our analysis shows that temperatures have been rising in recent years, with increases of 28°C in some years. The average annual temperature for the period studied at all stations is over 24°C, and the stations have relatively the same annual temperature values. In fact, these values are considered to be the lowest between 1983 and 1990 for the three observation stations over the entire time series (1983-2022). In addition to these low thermal values, we have average values recorded over several years with a value of 26°C. This average value recorded during certain months is the result of a period when we had not too abundant precipitation, around 1,200 mm to 1,800 mm. As for rising temperatures, all three stations recorded very high values, varying between 27°C and 28°C.

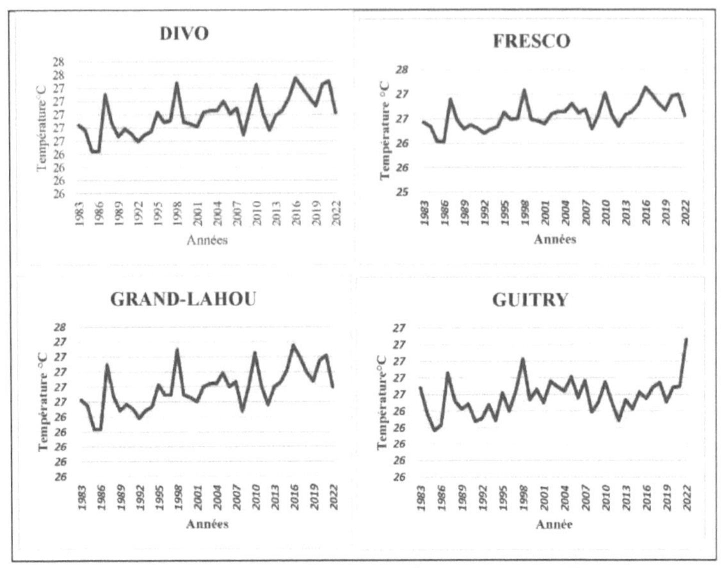

Data source: SODEXAM, NASA / Production: Kouadio Romaric 2021

Figure 4: Interannual thermal trends

1- 3- Determining climatic periods

Our study area is located in a sub-equatorial climate with a bimodal season. In fact, this season is animated by dry and wet periods, which are the different characteristics of the zone's climate. The Gaussen method, used to classify climatic periods, offers a structured perspective on rainfall variations throughout the year. By applying this method to the various stations, we can observe a detailed breakdown of the climatic periods from January to December. The table shows that January and December are classified as dry. This characterization suggests that during these months, precipitation is particularly low, marking the beginning and end of a dry period. This behavior is typical of a dry season that precedes and follows the rainy season. In February, however, the climate is average, indicating a transition between a dry period and the start of the wet season. Rainfall begins to increase slowly, signalling preparation for a wetter period. The months from March to September are then classified as wet, with variations ranging from average to very wet. More specifically, the months of March and April are initially wet, signifying an increase in precipitation that marks the start of the rainy season. Next, May and June are very wet, indicating periods of maximum rainfall, when precipitation is most intense. Finally, we have July and August, which continue in this wet phase, although the intensity may decrease slightly compared to previous months. September and October are also classified as wet, but tend towards a more average phase. This marks the end of the rainy season, with a gradual decrease in rainfall, signifying the transition to a drier period. Finally, November and December return to dry status, signaling the end of the rainy season and the return to a dry period. This transition indicates that precipitation is once again becoming scarce, completing the annual climatic cycle. This classification provides a better understanding of rainfall variations throughout the year, offering crucial information for resource management.

Table 2: Determination of Climatic Periods using the Gaussen Method

Divo												
	JAN	**FEV**	**MARCH**	**APRIL**	**MAY**	**JUNE**	**JUIL**	**AUGUST**	**SEVEN**	**OCT**	**NOV**	**DEC**
TEMP	2142,6	2262,2	2284,6	2263,2	2221	2122,2	2046,4	2023,8	2077	2140,8	2185,4	2144
PLUV	658,6	2226,2	4459,5	5743,4	7298	10611,8	4579,9	2322,4	4161,2	6253	4261,4	1854,6
PERIOD	Dry	Medium	Wet	Wet	Wet	Very damp	Wet	Medium	Wet	Wet	Wet	Dry
Fresco												
	JAN	**FEV**	**MARCH**	**APRIL**	**MAY**	**JUNE**	**JUIL**	**AUGUST**	**SEVEN**	**OCT**	**NOV**	**DEC**
TEMP	2126,2	2205,2	2242,6	2229,8	2179,8	2082,2	2013,6	2000,4	2035,8	2100	2159,6	2134,8
PLUV	897,3	2162,5	3503	5345,9	9210,7	14126,1	5464,7	2058,6	3617,7	5701,4	5126,9	1002,6
PERIOD	Dry	Medium	Wet	Wet	Wet	Very damp	Wet	Medium	Wet	Wet	Wet	Dry
Grand-Lahou												
	JAN	**FEV**	**MARCH**	**APRIL**	**MAY**	**JUNE**	**JUIL**	**AUGUST**	**SEVEN**	**OCT**	**NOV**	**DEC**
TEMP	2142,6	2229,2	2257,8	2254	2197,2	2098,8	2030,4	2007,6	2044,2	2117,4	2176,2	2150,4
PLUV	738,4	2029,1	3613,8	5591,5	9335,1	15182,3	5973,5	1586,7	2835,2	5403,8	5077,1	1365,3
PERIOD	Dry	Medium	Wet	Wet	Wet	damp	Wet	Medium	Wet	Wet	Wet	Dry
Guitry												
	JAN	**FEV**	**MARCH**	**APRIL**	**MAY**	**JUNE**	**JUIL**	**AUGUST**	**SEVEN**	**OCT**	**NOV**	**DEC**
TEMP	***2204,4***	***2224,6***	***2181,1***	***2126***	***2194,4***	***2013,7***	***2030,4***	***2007,6***	***2044,2***	***2117,4***	***2176,2***	***2167,2***
PLUV	***760,3***	***2007,7***	***3625***	***5591,5***	***9335,1***	***15182***	***5973,5***	***2051,7***	***3117,5***	***5322,4***	***5077,5***	***964,62***
PERIOD	Dry	Medium	Wet	Wet	Wet	Very damp	Wet	Medium	Wet	Wet	Wet	Dry

Data source: SODEXAM, NASA / Production: Kouadio Romaric 2021

2- Interannual evolution of hydrological conditions in the mangroves of the Tadio lagoon

2- 1- Interannual trends in potential evapotranspiration (PET) and actual evapotranspiration (AET)

2- 1-1-Evolution of potential evapotranspiration (ETP)

The figures illustrate annual potential evapotranspiration (PET) for the period 1983 to 2022. Potential evapotranspiration is a key indicator representing the maximum amount of water that could evaporate and be transpired by vegetation under optimal conditions. The graph shows that PTE has fluctuated over the years, but remains relatively stable around an average value of around 1000 mm/year. The lowest values were observed between 1985 and 1997, at around 1014 mm/year and 1020 mm/year respectively for the four analysis stations. Conversely, the highest values were recorded from 1998, 2016 and 2021, with values of around 1030 mm/year and 1050mm/year respectively. Indeed, these fluctuations are caused by various factors such as variations in temperature, humidity, solar radiation and wind speed, all of which influence evapotranspiration. Evapotranspiration is a key factor influencing the water availability in mangrove ecosystems. Variations in ETP can affect water salinity, which in turn can influence mangrove growth and distribution. ETP directly influences the amount of freshwater available to mangroves by affecting evaporation and transpiration. However, high PTE means that there is a high demand for water due to evaporation and transpiration. In this case, the supply of fresh water is insufficient, and mangroves can suffer from water stress, which can affect their growth and survival in certain areas. Conversely, a low PTE means that water demand is lower, which can promote freshwater retention in the soil and create more favorable conditions for mangroves. However, it is important to note that the impact of ETP on mangroves can vary according to other environmental factors such as rainfall, tides and the flow of surrounding watercourses. In our study area, high rainfall can compensate for high ETP by providing additional freshwater, especially in recent years.

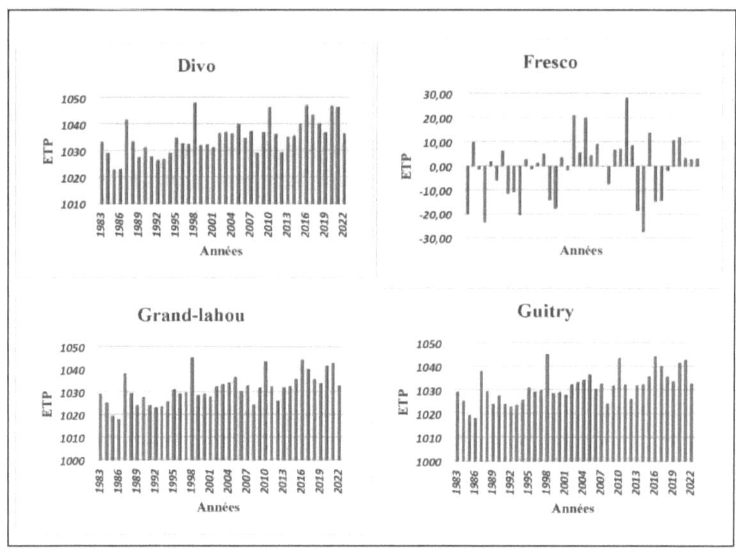

Data source: SODEXAM, NASA / Production: Kouadio Romaric 2021

Figure 5: Interannual trend in potential evapotranspiration

2- 1-2- Interannual trends in actual evapotranspiration (AET)

The figure below shows the evolution of the ETR (Evaporation-Transpiration-Recharge) index from 1983 to 2022 for four different stations. Examination of the data shows that the ETR index has fluctuated over the years. It has generally peaked in 1984, 2001, 2003, 2010, 1984, 2001 and 2014, exceeding 900 units. These peaks indicate periods of greater evaporation, transpiration and water recharge. Conversely, the ETR index experienced lows in 1986, 1990, 1991, 1992, 1998, 2012 and 2013, with values below 850 units. These troughs suggest periods of lower evaporation, transpiration and water recharge. Interestingly, since 2000, the ETR index appears to have been trending upwards, with values generally above 850 units, with the exception of 2012 and 2013. However, this trend is not constant and there are still fluctuations from one year to the next. The interpretation of this data may depend on the context. If we consider the ETR index to be an indicator of ecosystem health, years with a high ETR index could suggest conditions favorable to plant growth and water recharge. Conversely, years with a low REE index could indicate drier, less favorable conditions. However, it is important to note that the ETR index is influenced by many factors, such as precipitation, temperature, humidity, wind and solar radiation. Consequently, any interpretation of these data should take these factors and their evolution over time into account. In conclusion, this graph shows that the ETR index has fluctuated over recent decades, with a general upward trend since 2000. These fluctuations may have important implications for water resource management and ecosystem preservation in the region concerned.

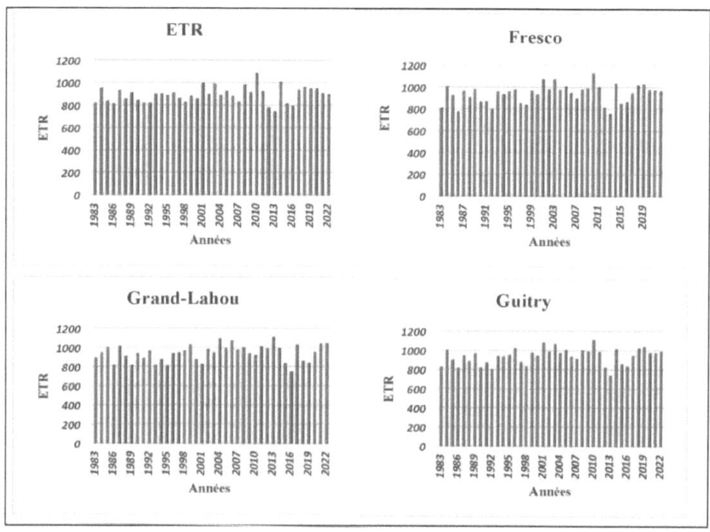

Data source: SODEXAM, NASA / Production: Kouadio Romaric 2021

Figure 6: Interannual trends in actual evapotranspiration

2- 2- Monthly trends in potential evapotranspiration (PET) and actual evapotranspiration (AET)

2- 2-1- Monthly trends in potential evapotranspiration (PET)

The figure shows the average monthly potential evapotranspiration (PTE), expressed in millimeters. First, PTE remains stable from January to March, with values ranging from 85 to 99 mm depending on the station. It then peaks in March, at around 99 mm, due to rising temperatures and spring sunshine. However, in April, a slight decrease is observed, although values remain high. Then, between May and August, ETP gradually decreases, from over 90 mm in May to less than 72 mm in August, due to lower precipitation and higher summer evaporation. In September, PTE rises slightly above 70 mm, but remains low compared with previous months. In October, it rises again to over 80 mm, in line with increased precipitation and lower temperatures. From November to December, PTE remained stable, fluctuating between 86 and 87 mm.

Thus, the ETP fluctuates annually according to the seasons and climatic conditions: it reaches its maximum values in spring and minimum in summer, influenced by precipitation and temperatures. These variations have an impact on mangroves, coastal ecosystems that require freshwater for their development and are sensitive to hydrological changes. Indeed, a high ETP increases the water demand of plants, which can induce water stress for mangroves in the event of a freshwater shortage, affecting their growth and survival. On the other hand, low ETP reduces this demand, favoring mangrove growth. However, mangroves adapt to seasonal fluctuations in PTE through various strategies, such as storing water in their leaves or stems and reducing transpiration to minimize water loss.

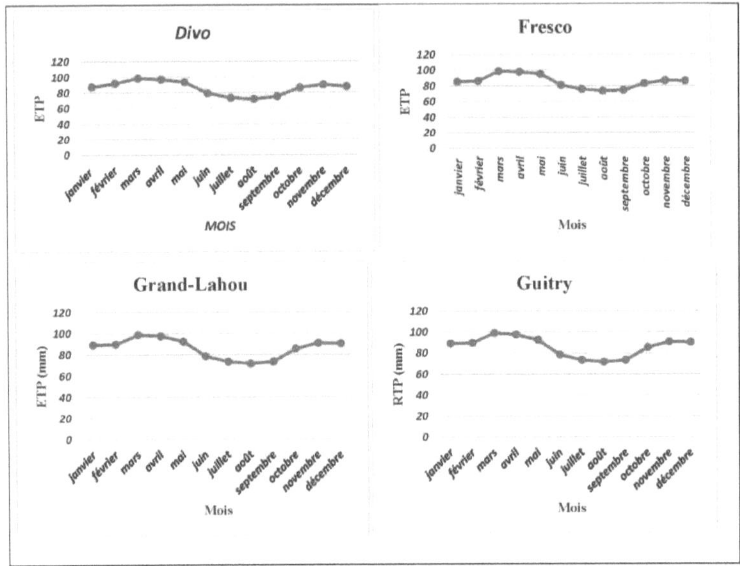

Data source: SODEXAM, NASA / Production: Kouadio Romaric 2021

Figure 7: Monthly trend in potential evapotranspiration

2- 2-2- Monthly trends in actual evapotranspiration (AET)

The chart below shows monthly variations in actual evapotranspiration. We can see that AET varies considerably from month to month. It is lowest in January (ranging from 15 to 23 mm) and highest in June (233.09 to 331 mm). This monthly variation in REE is mainly due to seasonal changes in temperature, humidity and precipitation. In general, REE rises as temperature and sunshine increase, and falls as temperature and sunshine decrease. This is why REE is highest in the hottest, sunniest months (March to June), and lowest in the coldest, sunniest months (December to February).

Changes in REE can have a significant impact on mangroves. High REE reduces the availability of fresh water for mangroves, particularly during dry months. This can lead to water stress for plants, which can affect their growth, reproduction and survival. Conversely, low REE can increase freshwater availability for mangroves, which can promote their growth and development. However, it is important to note that mangroves are adapted to seasonal variations in REE and have developed strategies to cope with periods of water stress. For example, some mangrove species can store water in their leaves or stems, while others can reduce transpiration to minimize water loss. However, mangroves are adapted to these variations and have developed strategies to cope with periods of water stress.

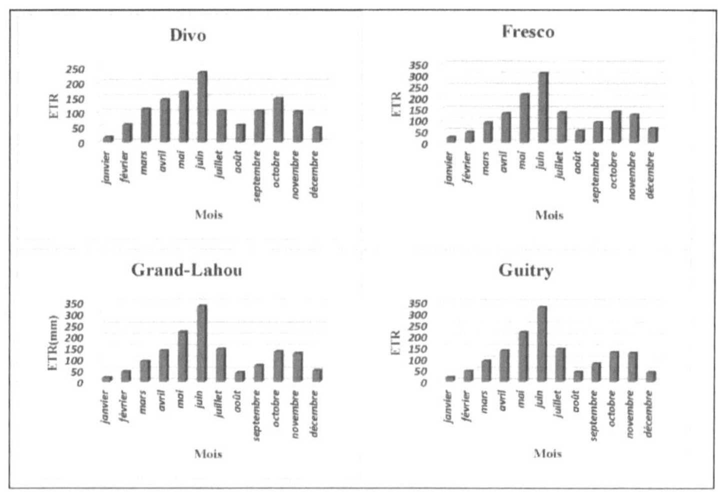

Data source: SODEXAM, NASA Production: Kouadio Romaric 2021

Figure 8: Monthly trend in actual evapotranspiration

II- Influence of Ocean Phenomena

1- Analysis of ocean indicators in Tadio lagoon

1- 1- Salinity and acidity of Tadio lagoon

Salinity is a key factor in the dynamics of mangroves in the Tadio lagoon, as it influences the distribution of the plant and animal species that inhabit this ecosystem. In fact, mangroves develop in areas where the water is brackish, i.e. where salinity is higher than in freshwater areas but lower than in marine areas. Water salinity in the Tadio lagoon is mainly due to the intrusion of seawater at high tide, but can also be influenced by other factors such as rainfall, river inputs and tides. From our investigations and analysis, it appears that the salinity of surface water is lower than that of water at depth, due to freshwater inflow from rivers and rainfall. In mangrove areas, water salinity can also vary according to the vegetation and shade provided by the trees. Mangrove roots can also affect water salinity by trapping sediments and reducing water circulation, which can increase salt concentration. The acidity of the water in Côte d'Ivoire's Tadio lagoon varies according to several factors, such as the decomposition of organic matter, the respiration of aquatic organisms, freshwater input and tides. In general, the pH of the water in mangroves is slightly acidic to neutral, with values between 6 and 8. However, in our case, the Tadio lagoon, water pH can be influenced by the presence of decomposing organic matter, which releases acids and reduces water pH. However, mangrove roots can also play an important role in regulating water pH by trapping sediments and promoting the growth of acid-neutralizing bacteria. According to the samples taken during our surveys, the pH of Tadio lagoon can vary between 6.5 and 7.5, with seasonal and spatial variations. However, these values can vary depending on local conditions and the year.

1- 2- Water temperature

Water temperature plays an important role in the dynamics of mangroves in the Tadio lagoon. Water temperature varies from 10°C to 47°C, depending on the level at which the measurement is taken. However, water temperature can vary according to depth and proximity to mangroves. In general, water temperature at the surface is higher than at depth, due to sun exposure and heat absorption. In mangrove areas, water temperature varies according to the vegetation and shade provided by the trees. Mangrove roots directly affect water temperature by reducing water velocity and promoting sediment accumulation, which in turn reduces the amount of sunlight entering the water and lowers temperature. In this mangrove area, the temperature varies between 20°C at the surface and 3°C with depth. However, it should be noted that water temperature at depth in the Tadio lagoon is more stable than at the surface, as it is less affected by variations in air temperature and sun exposure. However, water temperature at depth can also be affected by other factors, such as currents, tides and sources of fresh or salt water.

The surface water temperature of Tadio Lagoon varies according to a number of factors, including the season, time of day, weather and local environmental conditions. In the morning, before sunrise, the water temperature is lower, varying between 20°C and 25°C. This variation is also the result of surrounding activities. During the day, as the sun rises, the temperature starts to get very hot, and the sunshine on the surface of the lagoon gives it a temperature of between 25°C and 35°C. In some parts of the lagoon, temperatures can reach up to 47°C, especially near villages. In general, the surface water temperature in the Tadio lagoon is influenced by air temperatures and sun exposure.

1- 3- Phosphates and nitrates

Phosphate and nitrate concentrations in the Tadio lagoon vary according to several factors, such as nutrient inputs from rivers, human activities, decomposition of organic matter and biological processes. Phosphates and nitrates are essential nutrients for plant and algae growth in aquatic ecosystems. However, high concentrations of these nutrients lead to excessive algal blooms, with negative effects on water quality and the health of aquatic ecosystems. In the Tadio lagoon, phosphate and nitrate concentrations vary according to proximity to sources of nutrient input, such as rivers and surrounding villages. Human activities such as agriculture, livestock breeding and wastewater discharge also contribute to the increase in nutrient concentrations in the water. Phosphate and nitrate concentrations in the water of the Tadio lagoon can vary between 0.01 and 1.5 milligrams per liter for phosphates and between 0.1 and 10 milligrams per liter for nitrates, with seasonal and spatial variations. However, these values can vary depending on local conditions and the year.

2- Ocean currents and phenomena

2- 1- Marine currents and mangrove dynamics in Tadio lagoon

2- 1-1- Transport of nutrients and sediments by ocean currents

The transport of nutrients and sediments by marine currents is one of the main factors influencing the dynamics of mangroves in Tadio lagoon. Firstly, marine currents transport essential nutrients such as nitrogen, phosphorus and iron to the mangroves, which can promote their growth and development. Mangroves need these nutrients to maintain their productivity and health. Nutrients can be transported in dissolved form in water or as suspended particles. In addition, sediments are transported by ocean currents. Sediments transported by marine currents contribute to the formation of new soils and the expansion of mangroves. These sediments provide a stable substrate for mangrove growth and may also contain essential nutrients. However, excessive sediment also leads to siltation and mangrove death. In addition, we have Nutrient and Sediment Distribution. Marine currents help to distribute nutrients and sediments evenly throughout mangroves, which can contribute to uniform mangrove growth and development. Marine currents can also help transport nutrients and sediments to more remote areas of mangroves, which can help extend the reach of mangroves. Finally, marine currents directly affect water turbidity in the Tadio lagoon, which could have an impact on mangrove growth and survival. Turbid water can reduce the amount of sunlight reaching mangroves, which can affect their growth and development. Overall, the transport of nutrients and sediments by ocean currents is an important factor influencing the dynamics of mangroves in Tadio lagoon.

2- 2-2- Contribution to water salinity modification and coastal erosion

Sea currents play a decisive role in the salinity of the water in the Tadio lagoon, directly influencing mangrove dynamics. On the one hand, mangroves, which are adapted to saline environments, have variable tolerance levels depending on the species, which conditions their distribution and growth. On the other hand, these currents also bring freshwater into the lagoon via rainfall and rivers, helping to reduce the salinity required to maintain mangrove productivity. Seasonal fluctuations also influence salinity: the rainy season increases freshwater inflow, while the dry season intensifies evaporation, requiring mangroves to be more tolerant of these variations. Ultimately, these changes in salinity modify the structure of mangrove communities, impacting on their distribution and abundance.

With regard to coastal erosion, marine currents also contribute to the destabilization of riverbanks, resulting in soil loss that weakens the substrate essential for mangrove growth. In addition, sediment transport, often the cause of siltation, disrupts the stability and health of ecosystems. As waves increase in height and frequency, they exacerbate this erosion, while promoting the dispersal of propagules, enabling the colonization of new areas. Finally, the arrival of invasive species carried by these currents affects mangrove biodiversity. The figure below illustrates the annual variations in tidal range from 1983 to 2022, essential for analyzing the impact of tidal dynamics on these coastal ecosystems.

3- Evolution of Ocean Phenomena

3- 1- Interannual trends in oceanic phenomena

3- 1-1- Interannual tide gauge trends

The figure below illustrates annual variations in tidal range between 1983 and 2022, highlighting the inter-annual evolution in the Grand-Lahou region of southern Côte d'Ivoire. Tidal range, which expresses the difference between high and low tide, is a key parameter for understanding maritime dynamics and their impact on coastal zones. Overall, values fluctuate around 0.50, although there have been notable fluctuations over time. However, there has been an upward trend in recent decades.

At the beginning of the period, between 1983 and 1993, values mainly ranged between 0.47 and 0.55, characterizing a certain stability with limited variations. From 2006 onwards, a significant increase in tidal range is observed, as evidenced by values of 0.65 in 2006, 0.72 in 2007, then 0.78 in 2014. On the other hand, recent years show significant peaks, with tidal range reaching 0.80 in 2020, the highest of the period, while 2021 and 2019 show values of 0.79 and 0.77 respectively. In comparison, years prior to 2005 show less pronounced variations, revealing a relatively stable tidal range.

The increasing instability, particularly marked after 2005, could be associated with climatic and environmental factors. Some years, such as 1985 with a tidal range of 0.69 and 2014 with 0.78, show anomalies that indicate atypical increases. In short, the inter-annual analysis of tidal range between 1983 and 2022 suggests a significant increase, probably linked to phenomena such as climate change, rising sea levels or alterations in ocean currents. Consequently, this increased variability in recent years calls for sustained monitoring to better understand the underlying dynamics and their potential impacts on coastal ecosystems and human activities.

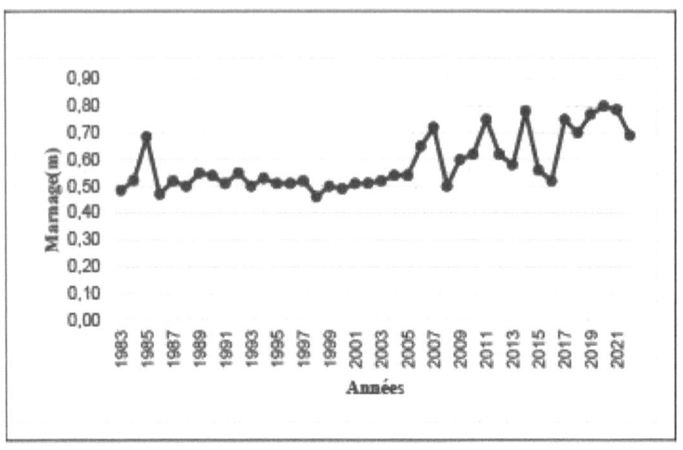

SIEREM 2021/ Project: Kouadio Romaric 2021

Figure 9: Interannual change in tidal range

3- 1-2- El Niño interannual trends

The following analysis focuses on the interannual evolution of the El Niño index in the Grand-Lahou region, Côte d'Ivoire, from 1983 to 2022. This index reflects sea surface temperature anomalies and their influence on local climatic conditions. Indeed, the years 1987, 1997, 2015 and 1983 are marked by significant positive anomalies, with indices of 15.32, 14.05, 17.53 and 5.76 respectively. These periods correspond to intense El Niño events, causing major climatic disturbances such as droughts and extreme rainfall variations. Conversely, the years 1988, 1999, 2000, 2011 and 2022 show significant negative anomalies, with indices of -9.73, -14.7, -10.03, -10.33 and -11.26 respectively. These values indicate La Niña events, generally associated with wetter conditions and flooding in the region. Other years, such as 1990, 1991, 1992, 1994, 2002, and 2004, show moderate positive anomalies with values oscillating between 3.01 and 7.76. These fluctuations indicate less intense El Niño periods, but still have an influence on the local climate. Years such as 1985, 1989, 2007, 2008, 2010, and 2016 show marked transitions with negative indices ranging from -7.18 to -10.33, demonstrating alternating El Niño and La Niña cycles. These transitions are crucial to understanding climate dynamics and their impact on ecosystems. Some years, such as 2005, 2006, 2014 and 2018, show near-zero anomalies, indicating relatively stable climatic conditions with no marked El Niño or La Niña events. In sum, it should be noted that the interannual evolution of the El Niño to index in the southern zone of Côte d'Ivoire reveals significant fluctuations between years of strong positive and negative anomalies. These variations are essential for understanding the impact of global climate cycles on local conditions. By identifying these trends, we can anticipate potential climate impacts and develop adaptation strategies for local populations and ecosystems. The data underline the importance of continuous monitoring of climate indices to better understand future environmental challenges.

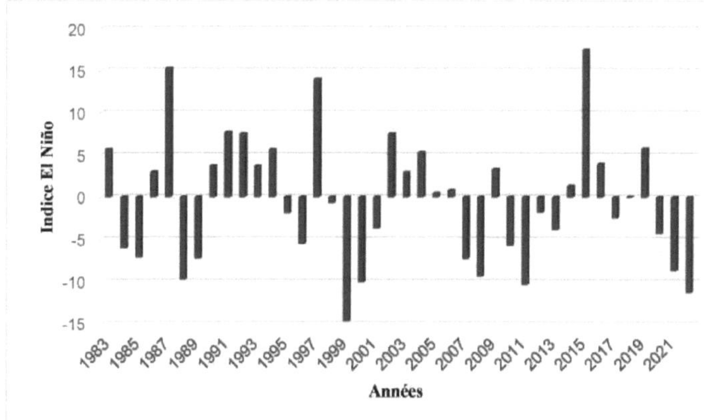

SIEREM 2021 / Designed by Kouadio Romaric 2021

Figure 10: Interannual evolution of the index]

3- 2- Monthly analysis of ocean phenomena

3- 2-1- Monthly tide gauge analysis

The analysis focuses on monthly tidal range trends at Grand-Lahou, a region in the south of Côte d'Ivoire. The months from January to April show relative stability, with a tidal range of 0.55 in January, 0.56 in February, then a slight decrease to 0.55 in March and April. This period shows little seasonal variation in tide levels, suggesting hydrological stability during the dry season. In May, the tidal range drops slightly to 0.54, continuing this trend in June with the same level, and in July reaching 0.53. This gradual decrease could be linked to increased rainfall and the onset of the rainy season, influencing tidal dynamics through increased river flows. August and September saw a slight return to 0.54. This relative stability could be explained by the peak of the rainy season, when tidal levels are stabilized by a constant supply of fresh water from precipitation and nearby rivers. In October, the tidal range rises again to

0.56, and remains at this level in November and December. This increase coincides with the end of the rainy season and the start of the dry season, when reduced precipitation allows a resumption of more pronounced tidal variations. Monthly tidal range trends at Grand-Lahou reveal a relative stability of tidal levels, with slight seasonal fluctuations. The dry period (January to April) shows relatively constant tide levels, while the rainy season (May to September) brings a slight decrease followed by stabilization. Finally, the resumption of more marked variations from October to December signals the transition to the dry season. These observations are essential for understanding local hydrological dynamics and planning coastal and resource management activities according to tidal variations.

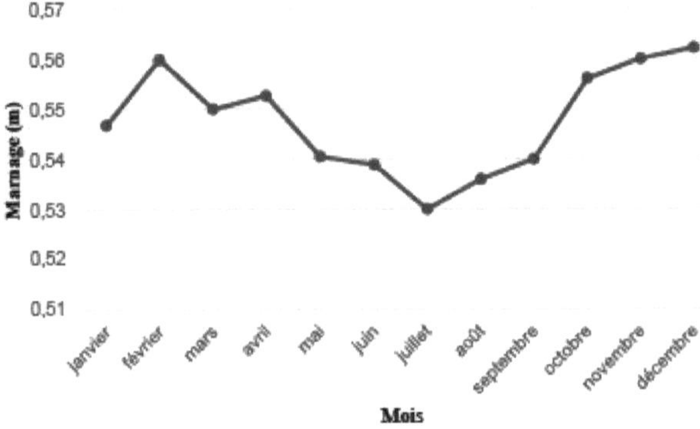

SIEREM 2021 / Designed by Kouadio Romaric 2021

Figure 11: Monthly change in tidal range

3- 2-2- Monthly analysis of the El Niño index

Analysis of the El Niño index at Grand-Lahou reveals significant monthly fluctuations. In January, the El Niño index is -0.81, slightly less negative in February at -0.69, and continues this trend in March at -0.4. This period shows a gradual decrease in the negative effects of the index, indicating a weakening of La Niña conditions. Between April and June, the El Niño index continues to show a less negative trend, with -0.35 in April, -0.33 in May and -0.24 in June. This evolution indicates a transition towards more neutral climatic conditions, marking an attenuation of negative thermal anomalies. In July, the index rises slightly to -0.27, before dropping again in August to -0.81. This sudden drop in August can be attributed to a temporary strengthening of La Niña conditions, characterized by lower sea surface temperatures. From September to December, the El Niño index shows a continuous and significant decline. In September, it reached -1.81, then -2.72 in October, -3.38 in November, and -3.84 in December. This period marks a marked intensification of La Niña conditions, with accentuated negative thermal anomalies. Ultimately, the first few months show an attenuation of negative conditions, suggesting a transitional phase towards more neutral conditions. However, from July onwards, a marked downward trend becomes apparent, culminating in December with very negative values, characterizing a strong La Niña influence. These variations have a direct impact on the local climate, influencing precipitation and temperatures, and must be taken into account in resource management.

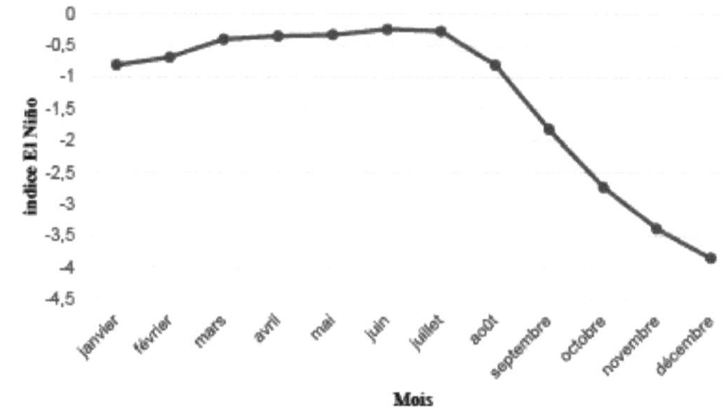

SIEREM 2021 / Designed by Kouadio Romaric 2021

Figure 12: Monthly evolution of El Niño indices

III- Influence of indices and oceanic phenomena

1- Impact of oceanic indices on mangrove dynamics in Tadio lagoon

1- 1- Impact of water salinity and acidity on mangrove evolution

The salinity and pH of the water in Tadio lagoon strongly influence mangrove dynamics. Firstly, salinity conditions the distribution and growth of mangrove species, each with a specific salt tolerance. Indeed, some, such as the red mangrove, adapt to high salinity environments, while others, such as the black mangrove, thrive better in less saline waters. In addition, salinity also affects marine fauna, many species of which depend on mangroves for their development, and which vary in abundance according to salt levels, enhancing the lagoon's biodiversity.

Secondly, the pH of the water affects the health of mangroves. When water is acidic (pH < 7), the availability of essential nutrients such as phosphorus and nitrogen decreases, limiting plant growth. Acidity also increases the toxicity of heavy metals such as lead, which can damage mangrove tissue. Conversely, alkaline water (pH > 7) reduces the availability of other elements such as iron, disrupting soil structure and root absorption capacity. However, mangroves develop adaptive mechanisms, adjusting the pH around their roots to favor their development despite acidic or alkaline variations in their environment.

1- 2- Impact of Water Temperature, Phosphate and Nitrate

The water temperature of the Tadio lagoon influences mangroves, affecting their growth, reproduction and composition. On the one hand, high temperatures can accelerate evaporation, causing water stress for plants and favoring more heat-resistant species, while low temperatures slow their growth and make them more vulnerable. In addition, temperature affects reproduction by influencing propagule germination and flowering, thus modifying the availability of propagules for the stand. It also interacts with other factors, such as salinity and storms, affecting ecosystem structure and health. In short, the impact of temperature on the mangroves of this lagoon is decisive, hence the need for in-depth local studies.

As for phosphates and nitrates, these nutrients have significant repercussions, notably by promoting the proliferation of algae, which limits the light available for mangrove photosynthesis. In addition, the decomposition of this excessive biomass releases toxins, upsetting the lagoon's ecological balance and reducing mangrove biodiversity. In addition, prolonged exposure to these nutrients can alter salinity, compromising the health of mangroves and making their leaves more vulnerable to disease. Conversely, a lack of nutrients can limit their development and reduce their photosynthetic capacity. Optimum management of phosphate and nitrate concentrations is therefore crucial to the sustainability of this ecosystem.

2- Impact of oceanic phenomena on mangrove dynamics in the Tadio Lagoon (southern Côte d'Ivoire)

2- 1- Impact of tidal evolution on mangrove dynamics in Tadio Lagoon (southern Côte d'Ivoire)

Tidal changes play a decisive role in the dynamics of mangrove ecosystems, particularly in the Tadio lagoon in southern Côte d'Ivoire. Variations in tidal levels influence hydrological conditions, directly impacting mangrove growth, composition and resilience. Firstly, tidal variations modify flood levels and salinity gradients. When tides are higher, mangrove areas are submerged for longer, favoring saltwater-tolerant species. Consequently, such a condition can lead to increased dominance of these species, altering the composition of the plant community. Secondly, lower tides expose mangrove roots to the air for prolonged periods. This exposure can increase water stress for less desiccation-tolerant plants, reducing their ability to survive and grow. As a result, species diversity may be reduced and ecosystem structure altered. Tidal changes also affect erosion and sedimentation processes. Stronger tides can lead to increased soil erosion, exposing mangrove roots and making young plants vulnerable. On the other hand, gentler tides can promote sedimentation, enabling the formation of new habitats suitable for mangrove colonization. However, mangroves possess adaptive mechanisms that enable them to respond to tidal variations. For example, some species develop stilt roots or pneumatophores to breathe better under conditions of prolonged inundation. In this way, the adaptive capacity of mangroves contributes to their resilience in the face of changing tidal levels. Finally, changes in mangrove dynamics due to tidal variations have significant ecological and socio-economic implications. Ecologically, they influence the biodiversity and productivity of coastal ecosystems. Socio-economically, local communities depend on mangroves for resources such as timber, fishing and protection against coastal erosion. Tidal variations can therefore have a direct impact on the livelihoods of local populations.

2- 2- Impact of ¡'El Niño' on Mangrove Dynamics in Tadio Lagoon (Southern Côte d'Ivoire)

The evolution of the El Niño phenomenon has major repercussions on coastal ecosystems, particularly on the mangroves of the Tadio lagoon in southern Côte d'Ivoire. The climatic variations associated with El Niño affect rainfall, temperatures and wind regimes, thus modifying the ecological conditions in which mangroves develop. Firstly, El Niño often leads to rainfall anomalies. In some regions, it causes prolonged droughts, reducing the availability of freshwater essential for mangrove growth.

This decrease in fresh water increases the salinity of the lagoon water, which can harm young plants and less salt-tolerant species, thus compromising the natural regeneration of mangroves. In addition, El Niño is often associated with temperature rises. Although mangroves are tolerant of temperature variations, they can suffer significant thermal stress during extreme heat events. This stress can reduce photosynthesis and plant growth, weakening the structure and function of the mangrove ecosystem. Furthermore, changes in wind regimes during El Niño periods can intensify coastal erosion. Stronger winds increase the force of waves, leading to increased erosion of the shores where mangroves are located. This erosion can damage mangrove roots, diminishing their stability and ability to reproduce. In addition, El Niño influences sea levels, sometimes causing them to rise temporarily. These fluctuations in sea level can excessively submerge mangroves, disrupting the tidal cycles to which they are adapted. Prolonged flooding can lead to plant mortality and altered species composition. Despite these challenges, mangroves are showing a degree of resilience to El Niño-induced climate variations. Some species possess adaptive mechanisms, such as the production of aerial roots to breathe better in the event of prolonged flooding. However, the increased frequency of El Niño events could overwhelm the adaptive capacity of these ecosystems, jeopardizing their long-term stability. However, the impacts of El Niño on the mangroves of the Tadio lagoon have important implications for conservation. Mangrove degradation affects not only local biodiversity but also the ecosystem services they provide, such as erosion protection and carbon storage. Appropriate management strategies are therefore needed to strengthen the resilience of these ecosystems in the face of climatic disturbances.

3- Mangrove growth, regeneration and mortality processes

3- 1- Sexual reproduction of mangroves

The sexual reproduction of mangroves in the Tadio lagoon is a complex and fascinating biological process that enables these remarkable trees to perpetuate themselves and maintain their vital presence in this unique coastal ecosystem. Indeed, the sexual reproduction of mangroves involves the production of flowers and seeds, as well as their dispersal and germination. Mangrove flowers are generally small and unspectacular, but they play an essential role in reproduction by attracting pollinators and enabling fertilization. In the Red Mangrove, sexual reproduction begins with the formation of flowers on the branches of the tree. After pollination, the flowers develop into elongated fruits, called propagules, which contain a single seed. The propagules are able to germinate and develop into a new tree while still attached to the parent tree. This reproductive strategy is particularly well-suited to conditions in the Tadio lagoon, where muddy, unstable soils make it difficult for seeds to germinate and take root. The Black Mangrove takes a slightly different approach to sexual reproduction. The flowers of this species grow in clusters on branches and, after pollination, give rise to capsules containing numerous seeds. Unlike the propagules of the Red Mangrove, the seeds of the Black Mangrove are released into the water and can float for several weeks before settling and germinating in a favorable environment. Finally, white mangrove is distinguished by its fragrant white flowers, which attract pollinating insects. After fertilization, the flowers turn into round fruits containing several seeds. White mangrove seeds are also dispersed by water, but have the particularity of germinating rapidly after exposure to air and sunlight. Sexual reproduction in the mangroves of the Tadio lagoon is an essential process for the maintenance and diversity of this coastal ecosystem.

The different reproductive strategies adopted by mangrove species illustrate their remarkable ability to adapt to difficult and constantly changing environmental conditions.

3- 2-Regeneration from mangrove stumps

Stump regeneration in mangroves is a remarkable biological process that enables these trees to recover and multiply after natural or anthropogenic disturbances. This ability to regenerate is essential for maintaining the health, resilience and biodiversity of mangrove ecosystems worldwide. Mangroves are trees adapted to life in salty and muddy coastal areas, and they face many environmental challenges such as storms, floods, droughts and tides. To overcome these constraints and ensure their survival, mangroves have developed various regeneration strategies, including stump regeneration. Stump regeneration, also known as stump sprouting or suckering, is a process whereby new shoots or stems emerge from the base of a damaged or cut tree. These new shoots, called sprouts, are genetically identical to the parent tree and can develop into new, independent trees. Stump regeneration is common in many mangrove species, including Red Mangrove (Rhizophora mangle), Black Mangrove (Avicennia germinans) and White Mangrove (Laguncularia racemosa). Red mangrove, for example, is known for its ability to produce abundant stump sprouts after being damaged or cut. Adventitious roots, which form on the lower parts of the tree stem, play a key role in this process. When the stem is cut or broken, the adventitious roots remain anchored in the soil, providing the nutrients and water necessary for the growth of new shoots. Black mangrove and white mangrove also have the capacity to regenerate from stumps, although the underlying mechanisms may vary slightly. In all cases, regeneration from stumps offers mangroves several ecological and evolutionary advantages. Firstly, it enables trees to recover quickly from disturbances, reducing the ecosystem's vulnerability to environmental stress and biological invasion. In addition, regeneration from stumps contributes to the vegetative propagation of mangroves, increasing the density and cover of mangrove forests. However, it is important to note that regeneration from stumps cannot always compensate for the loss of mangrove trees and forests. Large-scale disturbances, such as deforestation and coastal development, can exceed the regeneration capacity of mangroves, with serious ecological and socio-economic consequences. Thus, the conservation and restoration of mangrove ecosystems must be a top priority to preserve their beauty, biodiversity and invaluable ecosystem services. In short, regeneration from mangrove stumps is a fascinating and essential process that testifies to the resilience and adaptability of these exceptional trees. By understanding and appreciating the mechanisms of mangrove regeneration, we can strengthen our commitment to their protection and preservation for future generations.

Conclusion

The conclusion of this chapter highlights the significant effects of hydroclimatic changes on the mangroves of Tadio lagoon. Firstly, analysis of climatic parameters revealed monthly and interannual changes in rainfall and temperature, indicating a variation in climatic conditions that directly affects ecosystems. In parallel, hydrological conditions, illustrated by trends in potential and actual evapotranspiration, show a complex dynamic influenced by seasonal variations. Next, the influence of oceanic phenomena, such as salinity, acidity and marine currents, was examined. These factors not only modify the chemical composition of the water, but also impact the transport of nutrients and sediments, thus contributing to coastal erosion. The inter-annual evolution of these phenomena, notably tidal and El Niño variations, plays a crucial role in mangrove dynamics. Finally, oceanic indices influence biological processes such as mangrove reproduction and regeneration, underlining the importance of environmental stability for their survival. In sum, hydroclimatic and oceanic changes combined are exerting increasing pressure on the mangroves of Tadio lagoon, necessitating adapted management strategies to preserve these vital ecosystems.

Chapter 3: Anthropogenic Pressures and Mangrove Preservation Strategies

Introduction

The Tadio lagoon on the West African coast is an ecologically rich ecosystem, with mangroves helping to protect the coastline and maintain the environmental balance. However, urbanization and agricultural development around the lagoon have weakened this ecosystem. Urban expansion and infrastructure construction have often taken place at the expense of mangroves, polluting the water and soil with urban waste. In addition, intensive agriculture has led to the destruction of mangroves in order to free up space for cultivation, affecting both their surface area and water quality through the use of pesticides. Faced with this situation, preserving these mangroves is crucial, especially as they support the lives of local populations. In addition, the cultural and domestic use of mangrove resources by surrounding communities requires in-depth analysis. Thus, the first part explores traditional beliefs and practices linked to mangroves, while the second examines the economic and artisanal uses of these resources. Finally, it is important to foresee the prospects for future use, to ensure sustainable mangrove management for local generations. However, threats such as unsustainable fishing, aquaculture, deforestation, and wood and charcoal production are negatively impacting these ecosystems and the livelihoods of communities. With this in mind, it is necessary to promote conservation strategies, such as sustainable fishing practices, rotation of harvesting areas for timber, and reforestation methods. These initiatives should include training communities to strengthen their sustainable management capabilities, thereby supporting the ecological and economic resilience of the Tadio lagoon.

I - Impacts of Human Activities on Tadio Mangroves

1- Urbanization in the dynamics of the mangroves of the Tadio lagoon

1- 1- Urban expansion around the Tadio lagoon

Population growth around the Tadio lagoon has been gradual, fuelled by several factors. Firstly, the lagoon's rural location has attracted people seeking to escape the hustle and bustle of the cities to take advantage of the region's agricultural and livestock-raising opportunities. In addition, coastal erosion at Lahou-Pkanda has prompted some inhabitants to migrate to safer areas, contributing to population growth in the surrounding area. In addition, the lagoon itself is a magnet for fishing resources, river transport and commercial opportunities. In addition, the tranquility of the region and the tourist potential offered by villages such as Lozoua also contribute to this dynamic.

The urban exodus, meanwhile, has reinforced the area's demographic growth by attracting inhabitants in search of economic prospects and improved local infrastructures. This demographic increase, while economically advantageous, has generated increased environmental pressure, in particular degradation of lagoon water quality due to pollution and overexploitation of resources, affecting both biodiversity and human health. Thus, population growth around the Tadio lagoon, supported by a variety of factors, presents both beneficial effects for economic development and complex challenges for environmental sustainability.

1- 2- Dynamism of urban expansion in the Tadio lagoon area from 1990 to 2022

Our study area is a region rich in natural resources, which has experienced rapid population growth in recent decades. To better understand the changes that have taken place in this region, an analysis of the land use map in 1990, 2010 and 2022 was carried out. The results of this analysis revealed a remarkable change in the region's land use, with low occupancy in 1990, medium occupancy in 2010 and high human occupancy in 2022.

- **Urbanization still limited in 1990**

In 1990, urbanization around the Tadio lagoon was still at an embryonic stage, characterized by bare land and housing extending over 13 km^2 . This low level of urban expansion bears witness to a time when human pressure on the environment was limited. Human activities were mainly concentrated on traditional lifestyles, with minimal infrastructure. This could be explained by a relatively stable, sparsely populated area, economic activities centered on agriculture and fishing, and modest infrastructural development. Environmental impact, particularly on mangrove ecosystems, was probably minimal at this time, allowing the lagoon's natural environments to thrive without major disturbance.

Map 8: Land use in 1990

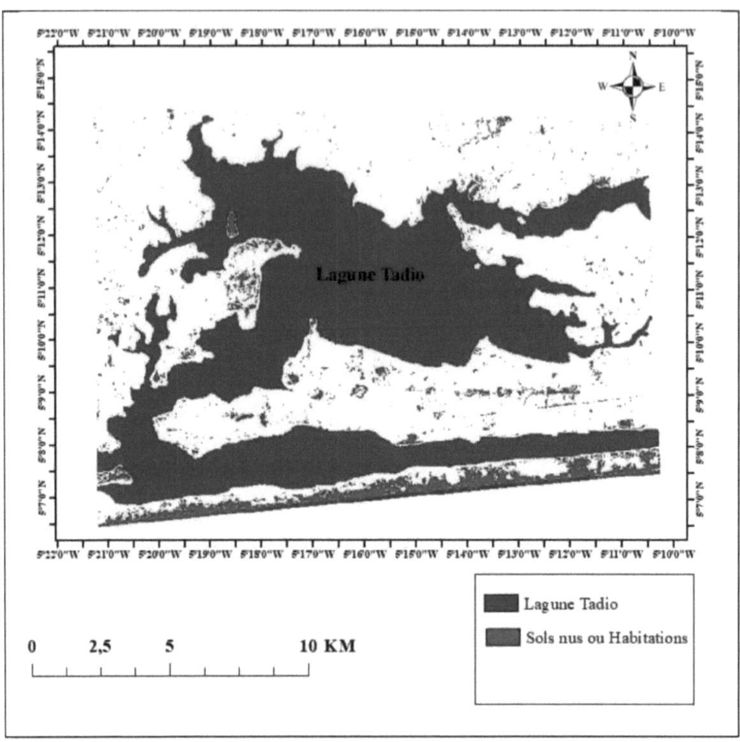

Source: Landsat 4-5-TM / Production: Kouadio Romaric 2022

➤ 2000 changing urbanization

In 2000, urbanization increased significantly, with the extension of residential areas and bare ground now reaching 28 km^2 . This growth in urbanization, almost twice as great as in 1990, indicates an acceleration of urban development in the region. This expansion can be attributed to several factors: demographic growth and a shift in economic activities requiring more built-up space. This phase marks a transition where human impact on the environment is becoming more palpable, probably affecting natural habitats more significantly. The management of land and natural resources is therefore becoming a crucial issue for the sustainability of the surrounding ecosystems, particularly for the mangroves that border the lagoon.

Map 9: Land use in 2000

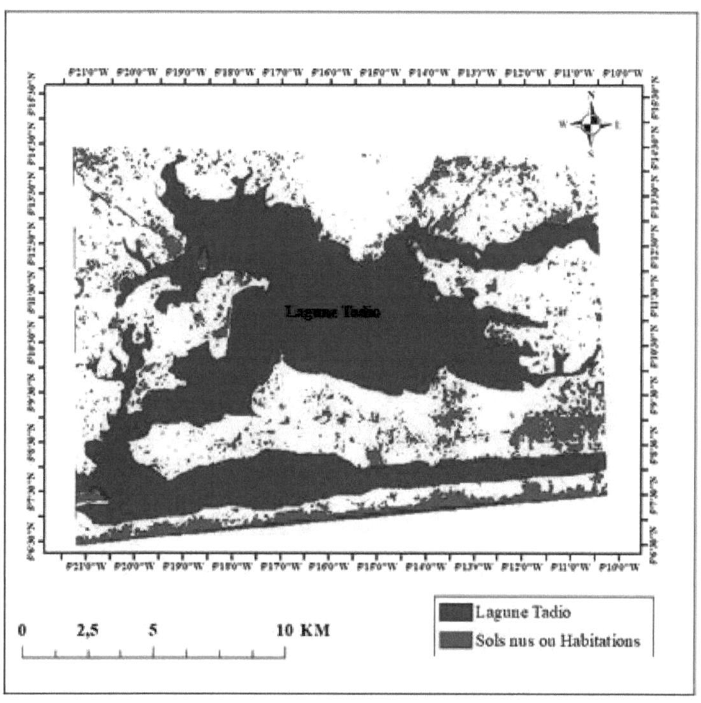

Source: Landsat 7-TM / Production: Kouadio Romaric 2022

- **2022 intensified urbanization**

Intensified urbanization in 2022, marked by an increase in bare land or housing (9894 hectares representing 5% of the area), is justified by several interconnected factors. Firstly, demographic growth has led to an increased need for housing, which in turn has led to the expansion of inhabited areas to the detriment of mangroves. Secondly, economic growth has favored infrastructure development, often carried out without sufficient consideration for the preservation of these ecosystems. Furthermore, the lack of strict conservation regulations has allowed mangroves to be exploited, facilitating construction and human activities. Finally, climate change is exacerbating the degradation of mangroves, making them more vulnerable to human pressures, thus indirectly contributing to their decline in favor of urbanization.

Map 10: Land use in 2022

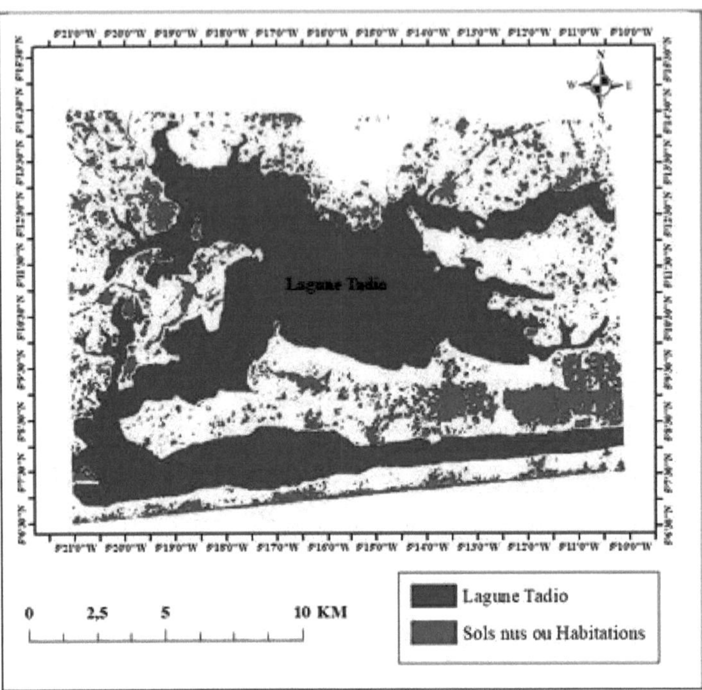

Source: Landsat 8-Oli / Production: Kouadio Romaric 2022

1- 3-Direct and indirect impacts of urbanization on mangroves

Urbanization and deforestation are two interconnected phenomena with significant impacts on the mangroves of the Tadio lagoon. Indeed, our field surveys have revealed the effects of urban expansion, which leads to a high demand for timber, resulting in the overexploitation of mangroves. In addition, waste and pollution generated by urbanization are degrading water and soil quality, compromising the health of these ecosystems.

In addition, deforestation is a crucial factor in habitat loss, with mangrove areas converted to farmland or exploited for timber. This leads to habitat fragmentation, with adverse consequences for biodiversity and the local communities that depend on these resources.

As far as water pollution is concerned, urbanization increases waste and sewage, causing blockages and harmful algal blooms, which disrupt aquatic ecosystems. This pollution affects biodiversity and the resources available to local communities, who often depend on fishing and agriculture.

Urbanization also disturbs flora and fauna, leading to a loss of biodiversity due to the fragmentation of natural habitats. Noise and light pollution, as well as atmospheric pollution, also affect the behavior of animal and plant species.

Finally, we found that urbanization modifies the hydrological regime and salinity of mangrove ecosystems. Construction disrupts natural water flows, leading to stagnation and reduced freshwater input, which, combined with increased demand for domestic and agricultural needs, causes saltwater intrusion. These disruptions alter ecological processes and compromise the health of species living in these ecosystems, thus affecting water and soil quality.

Photo 2: Degraded mangrove area

2- Influence of Human Activities on Tadio Mangroves

2- 1- Agricultural development in the Tadio lagoon region

The Tadio lagoon area features a diversity of crops, influenced by local ecological and economic conditions. On the one hand, food crops dominate, including rice in flood-prone areas, while cassava and yams are found on higher ground. Plantain and maize are grown as fallows or in association. Vegetable crops such as tomatoes and eggplants thrive on irrigated land. Some cash crops, though less widespread, include rubber, cocoa and oil palm, grown on higher ground.

In terms of farming practices, the region is predominantly oriented towards traditional family farming, characterized by the use of hand tools. However, recent initiatives have introduced agroecological techniques, promoting environmental sustainability through crop rotation and the use of compost. As a result, these new practices aim to improve productivity while preserving the lagoon's biodiversity.

As for the impact of agriculture on land use, it appears to have a significant influence, particularly on mangroves. These ecosystems, valuable for their coastal protection and carbon sequestration services, are threatened by agricultural expansion. The expansion of crops such as maize and manioc, as well as cash crop plantations, is leading to land clearance that is reducing forest cover. As a result, intensively farmed mangrove soils are losing their fertility, posing a risk to the livelihoods of local populations. These populations depend on mangroves for fishing and subsistence farming. In addition, monoculture can affect food security by limiting the diversity of local diets.

2- 2- Poverty and overexploitation of resources

Local communities in the Tadio lagoon rely heavily on mangroves for food, income and protection against natural disasters. These ecosystems provide essential fish, shellfish and molluscs for local consumption and trade, while offering a source of income through fishing and the collection of non-timber forest products. In addition, mangrove roots stabilize soils, reduce erosion and protect against storms and flooding. These forests also have a cultural value, being places of worship for the local populations.

However, this dependence leads to unsustainable fishing and aquaculture practices, such as overfishing and intensive shrimp farming, which affect biodiversity and the ability of mangroves to provide ecosystem services, resulting in water and soil pollution. On the other hand, the adoption of sustainable methods, such as gillnetting or integrated aquaculture, could preserve these resources.

In addition, overfishing and the destruction of breeding habitats in mangroves compromise the regeneration of marine species, upsetting the ecological balance and reducing biodiversity. Excessive shellfish harvesting also has similar repercussions, threatening the sustainability of these ecosystems. With this in mind, the introduction of quotas and environmentally-friendly harvesting practices would help minimize these impacts. Finally, encouraging local consumption and sustainable sales channels would strengthen the local economy while reducing environmental impacts, thus promoting balanced management of the natural resources of the Tadio lagoon mangroves.

2- 3- The direct effects of overexploitation on the local community

Firstly, the over-exploitation of mangrove resources in the Tadio lagoon compromises their productivity, reducing the income of local communities. These ecosystems are home to a variety of marine species essential to local subsistence, but excessive fishing and harvesting disrupt food chains, reducing overall productivity. In addition, excessive cutting of trees for timber reduces the capacity of mangroves to store carbon and provide nutrients for marine species, indirectly impacting income from fishing and shellfish harvesting.

Secondly, overexploitation leads to a loss of natural protection against storms and waves, endangering surrounding land and property. The roots and trunks of mangroves, by forming a natural barrier, normally attenuate the force of waves, protecting coastal areas against erosion. However, the decreasing density of mangroves weakens this protection, increasing the risk of flooding and degradation. Finally, this degradation has economic and social consequences, as erosion and natural disasters lead to loss of livelihoods for fishermen, harvesters and landowners, threatening the security of local communities.

3- How the local population uses the mangroves of the Tadio lagoon

3- 1- Domestic uses of mangroves in Tadio lagoon

The use of mangrove wood in coastal areas is of vital importance to local populations, who make use of it for both construction and heating. Indeed, communities exploit this material for traditional architectural techniques adapted to the specific conditions of coastal regions. Thanks to its strength and durability - conferred by natural tannins - mangrove wood is ideal for structures capable of withstanding weathering and marine aggression. What's more, its high density and low moisture content make it an efficient fuel for cooking and heating, helping to reduce wood consumption. However, unregulated logging can weaken the ecosystem of the Tadio lagoon.

At the same time, mangrove aquaculture, also known as "marine forestry", offers encouraging prospects for food security and local economic development. It contributes to enriching local diets and provides a source of income through the sale of aquaculture products. Mangrove leaves and bark play a crucial role in this activity, providing bait and nutrients for fish farms, as well as serving as material for fishing equipment. In this way, aquaculture promotes sustainable mangrove management, enhancing the value of this ecosystem while meeting the needs of local communities.

3- 2- Artisanal uses of mangroves in Tadio lagoon

In the coastal regions of Côte d'Ivoire, the mangroves of the Tadio lagoon provide a precious wood resource, used by local craftsmen to create a wide variety of products, such as furniture and sculptures. Mangrove wood, appreciated for its durability and resistance, is transformed using traditional techniques that reveal the craftsmanship of the carvers and joiners. In addition, the creation of unique furniture, adorned with natural veins and nuances, illustrates the skill of these craftsmen. As for the sculptures, they embody a folk art featuring characters, animals and abstract motifs, while promoting local culture through exhibitions in Côte d'Ivoire and elsewhere.

In addition to wood products, the region's artisans also exploit the mangroves' non-wood resources, such as leaves for basketry and bark for dyeing, testifying to the richness of the local craft heritage. Leaves, for example, are transformed into utilitarian and decorative objects using weaving techniques, while bark offers a palette of natural hues, used to dye textiles. In addition, the roots, with their distinctive shapes, are used to make jewelry and decorations, as part of the local craft industry's tribute to the coastal landscape.

On the medicinal front, mangroves are also renowned for their therapeutic properties. Species such as *Rhizophora mangle and Avicennia germinans* are used to treat a variety of ailments, thanks to their leaves, barks and roots with proven benefits. Finally, the preparation and administration of these remedies vary according to symptoms, including decoctions, powders and macerations, which exploit the plants' active principles in a targeted way. Decoctions, for example, are used to treat digestive disorders, while powders and macerations are used to treat muscular pains and inflammations, adapting to users' needs and perpetuating traditional knowledge.

II- strategies for preserving the mangroves of the Tadio lagoon

1- Influence of cultural factors on the use of mangroves in Tadio lagoon

1- 1- Traditional beliefs and practices related to mangroves

The mangroves of the Tadio lagoon are ecosystems that provide many cultural benefits to local communities for local rituals and ceremonies. Indeed, these mangroves are sacred places for many local communities. Indigenous peoples and coastal communities have a close relationship with mangroves, which are often considered places of birth, eternal rest and communication with ancestors. Mangroves are also associated with spirits and deities, who are venerated and invoked during rituals and ceremonies. Mangroves also provide many natural resources used in local rituals and ceremonies. The leaves, bark, roots and fruits of mangrove trees are used to make traditional medicines, dyes, perfumes and incense. Mangrove shellfish, crustaceans and fish are also used in offerings and ritual meals. Mangroves are also places where traditional knowledge and culture are passed on. Rituals and ceremonies are often opportunities for community members to come together, share stories and legends, and pass on knowledge about medicinal plants, sustainable fishing and harvesting practices, and the relationship between humans and nature. Finally, the role of mangroves in local rituals and ceremonies can contribute to the conservation and restoration of these vital ecosystems. Local communities with a close relationship with mangroves are often more likely to protect and manage them sustainably. Rituals and ceremonies can also strengthen social ties and community cohesion, which can facilitate the implementation of mangrove conservation and restoration projects.

In parallel, we have taboos and prohibitions, which are cultural and social practices that govern the use and exploitation of natural resources in many communities around the world. The mangroves of the Tadio lagoon are no exception to this rule. Taboos and prohibitions play an important role in the conservation and sustainable management of the mangroves of the Tadio lagoon. Local communities often have in-depth knowledge of their natural environment and the resources available there. Taboos and prohibitions can therefore reflect this knowledge and be used to protect the most vulnerable species and habitats. For example, some communities in the Tadio lagoon have taboos against cutting down certain mangrove trees, as they are considered sacred or as harboring spirits. These taboos help to preserve the biodiversity of the mangroves and maintain the health of the ecosystem. In addition, taboos and prohibitions help to reduce conflicts and tensions linked to the use and exploitation of mangroves in the Tadio lagoon. Local communities often have different interests and needs when it comes to mangrove use. Taboos and prohibitions also help to establish clear rules and norms for resource use, which can reduce conflicts and tensions between different user groups. For example, some communities in the Tadio lagoon have taboos against fishing in certain mangrove areas, as these areas are considered spawning grounds or nurseries for fish. These taboos help to protect fish resources and reduce conflicts between fishermen and other mangrove users. Finally, taboos and prohibitions are often misunderstood or misapplied, leading to unsustainable resource use practices, and are influenced by external factors such as urbanization and climate change, weakening their effectiveness and relevance.

1- 2- Traditional knowledge of mangrove biodiversity

The mangroves of the Tadio lagoon have an essential cultural and spiritual value for local communities, who practice offerings and rituals for protection and fertility. These ancestral traditions, integrated into their daily lives, aim to solicit the spirits' blessings to ensure the fertility of the soil and the productivity of fishery resources. Offerings include various gifts, accompanied by prayers and songs, while rituals, organized before the rainy and fishing seasons, involve dances and blessings to preserve the mangroves from malevolent spirits. These practices, guided by spiritual leaders, strengthen the bond with the natural environment and perpetuate cultural values, despite the growing influence of modernization and globalization, which is undermining young people's attachment to these traditions.

In addition, customary prohibitions define the periods and zones in which mangroves can be exploited to guarantee their sustainability. These restrictions, perceived as sacred, aim to maintain ecological balance by limiting the exploitation of resources to certain seasons and parts of the mangroves. In this way, they ensure the preservation of biodiversity and strengthen social ties within communities. However, the influence of Western cultures and demographic pressure are making their application more complex, threatening the transmission of these ancestral practices.

1- 3- Issues and challenges related to preserving the cultural uses of mangroves in the Tadio lagoon

The mangroves of the Tadio lagoon are among the ecosystems threatened by various anthropogenic pressures, such as deforestation, pollution and climate change. Preserving the cultural uses of mangroves is therefore a major challenge for the conservation of these ecosystems and for maintaining the traditional practices and knowledge of local communities. Firstly, the cultural uses of mangroves are closely linked to the identity and social cohesion of local communities. Mangroves are places for ritual practices, ceremonies and the transmission of knowledge and values between generations. The loss or degradation of these places can therefore have significant impacts on the identity and social cohesion of communities, and on their ability to maintain their traditional practices and knowledge. Secondly, the cultural uses of mangroves are also closely linked to food security and the local economy of communities. Mangroves are important fishing and harvesting areas for local communities, providing them with fishery resources and non-timber forest products such as honey, leaves and bark. The loss or degradation of these areas can therefore have significant impacts on the food security and local economy of communities, and on their ability to maintain the traditional practices and knowledge associated with these activities. However, preserving the cultural uses of mangroves is a major challenge, due to the various anthropogenic pressures that threaten these ecosystems. Deforestation, pollution and climate change are all factors that can have significant impacts on the biodiversity and natural resources of mangroves, and therefore on the cultural uses of local communities. Moreover, preserving the cultural uses of mangroves is also a challenge due to the low recognition and valuation of these uses and traditional knowledge by economic players and political decision-makers. The cultural uses of mangroves are often considered secondary to economic uses, such as shrimp or firewood production, and are therefore little taken into account in development policies and projects. Finally, preserving the cultural uses of mangroves is a challenge due to the low involvement and participation of local communities in the decision-making and management processes of these ecosystems. Local communities are often poorly consulted and involved in mangrove development and conservation projects, which can have a significant impact on their ability to maintain their traditional practices and knowledge linked to these ecosystems.

2- Environmental and regulatory framework for the preservation of mangroves in the Tadio lagoon

2- 1- Côte d'Ivoire Environmental Code and its provisions on mangrove preservation

The general provisions of Côte d'Ivoire's Environmental Code (Law no. 96-766 of October 3, 1996) establish a legal framework for the sustainable preservation of ecosystems. On the one hand, it stipulates that any degradation of natural environments, whether terrestrial or aquatic, must be avoided, with specific measures for forests, wetlands, mangroves and coasts. Secondly, any human activity with a potential impact on these ecosystems requires prior assessment (Article 39) to minimize damage. In addition, the fight against pollution (Article 57) extends to all areas to limit contamination of air, water and soil, particularly in areas rich in biodiversity. In addition, the law prohibits the destruction of protected species (Article 72) except with exceptional scientific authorization, and requires the rehabilitation of degraded ecosystems (Article 85). In short, this code lays down rigorous rules to guarantee ecological management for the benefit of present and future generations.

Mangroves benefit from enhanced protection under the Environment Code, including specific safeguards. Firstly, the articles of Chapter III strictly regulate coastal activities and require an environmental assessment (art. 39). Chapter IV prohibits all deforestation or degradation, and article 88 regulates economic projects, requiring impact studies to avoid alterations. Complementarily, the National Action Plan for the Environment (PNAE) and Decree no. 2015-537 reinforce these provisions by incorporating guidelines for restoring and conserving mangroves in collaboration with local communities.

The Code also provides for the participation of local populations in mangrove management, by involving these communities in the monitoring, protection and sustainable management of these sensitive ecosystems (Article 65). This involvement aims to strengthen local capacities and enhance traditional knowledge, for an ecological and economically viable management of mangroves, while respecting their biodiversity and importance for future generations.

2- 2- Other national laws and regulations relevant to mangrove management

National legislation and regulations on coastal zone management in Côte d'Ivoire provide a comprehensive legal framework for the protection of vulnerable ecosystems. On the one hand, Decree no. 2015-537 requires environmental assessments for projects affecting coastal zones, aimed at limiting the impact of agriculture, urbanization and the exploitation of natural resources. At the same time, Law no. 2015-532, or the Sea Code, provides a framework for the exploitation of marine and coastal resources, imposing sanctions against pollution and overfishing. In addition, the National Plan for Integrated Coastal Zone Management (PNGIZC) coordinates sustainable management in partnership with international bodies, encouraging local participation. Finally, the 2014 Environmental Code strengthens the protection of coastal ecosystems in the face of climate change.

As for the texts relating to the fight against climate change, the Environment Code, established by Law no. 96-766, introduces standards for the management of natural resources and the reduction of emissions. Then, Law no. 2014-390 on Energy Transition promotes renewable energies and energy efficiency, limiting greenhouse gas emissions. In addition, Law no. 2019-700 protects forests and prevents deforestation, thus contributing to carbon sequestration. Similarly, the National Strategy for Sustainable Development (Stratégie Nationale de Développement Durable - SNDD) aims for growth in harmony with environmental preservation, while the National Climate Change Adaptation Plan (Plan National d'Adaptation au Changement Climatique - PNA) proposes targeted measures for vulnerable sectors. Finally, the technical standards governing environmental impact assessments ensure that projects comply with climate and environmental criteria, by imposing corrective measures to reduce any negative effects.

III - Mechanism for preserving mangroves in Tadio lagoon and outlook

1- Tadio Lagoon mangrove preservation mechanism

1- 1- Community actions to preserve the mangroves of the Tadio lagoon

The mangroves of the Tadio lagoon are crucial ecosystems for biodiversity and local communities. On the one hand, capacity-building for local populations, through training in sustainable fishing, waste management and mangrove restoration, enables them to play an active role in preserving these ecosystems. This process, often supported by NGOs and local agencies, involves integrating sustainable practices and raising awareness of the benefits of mangroves for biodiversity and local resources. On the other hand, empowering communities, notably through the creation of community management committees and the recognition of customary land rights, promotes informed decision-making and encourages sustainable economic activities such as ecotourism. This approach also strengthens local governance and reduces conflicts over natural resource management.

However, sustainable aquaculture and fishing practices are essential to preserve the biodiversity of the Tadio lagoon. The establishment of protected fishing zones and the training of fishermen in less destructive techniques, such as the use of adapted nets, limit the overexploitation of resources. In aquaculture, the integration of eco-responsible systems, including the use of local species and the reduction of pollution, helps to maintain an ecological balance. However, the sustainability of these practices requires that they be integrated into global development strategies, simultaneously addressing issues such as poverty and climate change.

Last but not least, training and raising community awareness of mangrove preservation issues are crucial. The implementation of training programs in sustainable management, environmental restoration and ecosystem monitoring, designed in consultation with communities, helps to anchor sustainable practices. Moreover, raising local populations' awareness of the economic and environmental benefits of mangroves strengthens their commitment to preserving this unique environment. To complement this, the capacities of local and national authorities need to be developed, particularly in terms of implementing mangrove management and ecosystem monitoring policies, in order to establish a participatory and inclusive approach involving all stakeholders.

1- 2- Measures to protect mangrove forests from deforestation and logging

Preserving the mangroves of the Tadio lagoon is crucial for biodiversity, local livelihoods and the fight against climate change. Firstly, unsustainable agriculture and increasing urbanization threaten this ecosystem. Intensive farming practices, such as slash-and-burn or chemical use, degrade vegetation cover, promote erosion and pollute water. To remedy this, agroforestry and organic farming, which protect the soil and limit pollution, should be encouraged.

Secondly, rapid urbanization is encroaching on mangroves, leading to habitat loss and ecological fragmentation. Sustainable urban planning, focused on protecting these areas, and restoration initiatives, including the creation of nature reserves, will help preserve mangroves.

Furthermore, in the face of illegal logging, systems of surveillance and sanctions in the surrounding villages, under the impetus of local chiefs and national authorities, would strengthen conservation by actively involving the population.

Finally, to counter water pollution and soil degradation, strategies to treat wastewater and reduce the use of agrochemicals are essential. Participatory and transparent governance, involving all stakeholders and raising the awareness of local communities, is essential for the integrated and sustainable management of the mangroves of the Tadio lagoon.

2- Future prospects and recommendations for the mangroves of Tadio lagoon

2- I-Perspectives and opportunities for the sustainable preservation and development of mangroves in the Tadio lagoon

To preserve and restore mangroves, national and international policies, strategies and regulatory frameworks have been developed. Several countries have adopted national measures, such as Côte d'Ivoire's sustainable strategy, updated in 2020, to improve governance and restore degraded areas. At international level, conventions such as the Convention on Biological Diversity and the UNFCCC also promote the preservation of these ecosystems, recognizing their role in combating climate change.

The development of tools and approaches for the integrated management and ecological restoration of mangroves has also been stepped up. The use of geographic information systems (GIS) and remote sensing facilitates mangrove monitoring. In Côte d'Ivoire, marine and coastal spatial planning (MSCP) is underway for areas such as the Tadio lagoon. Innovative restoration techniques, such as mangrove planting and sediment retention structures, are showing promise in strengthening the resilience of mangroves in the face of ecological challenges.

Finally, building the capacity of local communities is crucial. Training and awareness-raising programs help them understand the importance of mangroves and adopt sustainable practices. Access to economic resources and greater participation in decision-making also strengthen their resilience in the face of environmental and social change. In short, concerted efforts by local and international players are essential to ensure the sustainability of these precious ecosystems.

2- 2- Recommendations for action and research for the sustainable preservation and development of mangroves in the Tadio lagoon

Firstly, regulatory frameworks play a crucial role in protecting mangroves and ensuring their sustainable use, by regulating human activities such as logging, fishing and coastal development. In addition to the creation of protected areas, their effective implementation is essential, requiring adequate resources for regulatory bodies, as well as rigorous monitoring. Local communities and other stakeholders are also involved in this process, reporting violations and supporting enforcement efforts.

In addition, sustainable management plans reinforce this protection by identifying objectives, threats and opportunities for mangroves. They provide for concrete measures as well as monitoring and evaluation mechanisms, which in turn require effective implementation to ensure their effectiveness. The participation of local communities is essential to evaluate the proposed actions and adjust them as required.

From a complementary perspective, the development of local preservation and restoration initiatives and projects is essential. Adapted to the conditions of local communities and ecosystems, these projects are optimized by the participation of stakeholders (NGOs, companies, governments). In addition, modern technologies such as remote sensing and computer modeling support the mapping of priority areas, improving understanding of ecosystems and the management of initiatives.

Finally, to meet the complex challenges facing mangroves, an integrated, collaborative approach is essential. This involves creating synergies between various sectors such as agriculture, fisheries and coastal development, and building the capacities of local communities to increase their resilience in the face of change. In short, the strengthening and effective implementation of regulatory frameworks, combined with local projects, constitute a coherent approach to sustainable mangrove management.

2- 3- Recommendations for the use of mangrove species in modern medicine

To promote research into the medicinal properties of mangrove species, we need to set up targeted research programs and foster close collaboration between researchers, traditional medicine practitioners and local communities. These programs aim not only to identify mangrove species with potential medical applications, but also to develop local capacities through training and research facilities. At the same time, they encourage sustainable management of natural resources, integrating responsible harvesting practices and protecting vulnerable species.

At the same time, communicating and popularizing research findings remains crucial to raising awareness among a variety of audiences, including healthcare professionals, policy-makers and local communities. Different communication strategies adapted to each group (scientific publications for professionals, educational campaigns for communities, reports for decision-makers) facilitate understanding and adoption of sustainable practices.

To strengthen exchanges between scientists, traditional practitioners, policy-makers and communities, the creation of dialogue platforms and working groups is essential. These spaces support a better understanding of challenges and encourage common solutions, while valorizing traditional knowledge. An integrated approach, involving various sectors such as agriculture, fisheries and coastal development, is essential to maximize synergies and adapt management strategies to current environmental and social challenges.

Conclusion

In short, this chapter has highlighted the multiple pressures exerted by human activities on the mangroves of the Tadio lagoon. Firstly, increasing urbanization, illustrated by urban expansion between 1990 and 2022, has generated significant impacts, both direct and indirect, on this fragile ecosystem. In addition, agricultural development, coupled with poverty and overexploitation of resources, has exacerbated the challenges facing mangroves, with adverse consequences for the local communities that depend on them. Nevertheless, preservation initiatives are emerging, based on cultural factors and environmental regulations. Traditional beliefs and local knowledge of biodiversity play a crucial role in the sustainable management of these ecosystems. In addition, the legislative framework, such as Côte d'Ivoire's Environmental Code, is an essential instrument for the protection of mangroves. In conclusion, the preservation mechanisms already in place, notably community actions and protection measures, bear witness to a growing awareness. For the future, it is imperative to promote sustainable strategies, including recommendations on the use of mangroves in modern medicine, to ensure their enhancement and preservation. In this way, an integrated and participatory approach could guarantee the sustainability of the mangroves of the Tadio lagoon in the face of growing anthropogenic pressures.

REFERENCE BIBLIOGRAPHIQUE

Tomlinson, P. B. (1986). *The Botany of Mangroves.* Cambridge University Press.

Alongi, D. M. (2009). *The Energetics of Mangrove Forests.* Springer.

Duke, N. C. (1992). *Mangrove Floristics and Biogeography.* In *Tropical Mangrove Ecosystems,* eds. A. I. Robertson and D. M. Alongi, 63-100. Washington, DC: American Geophysical Union.

Barbier, E. B., Hacker, S. D., Kennedy, C., Koch, E. W., Stier, A. C., & Silliman, B. R. (2011). The value of estuarine and coastal ecosystem services. *Ecological Monographs*, 81(2), 169-193.

FAO (2007). *The World's Mangroves 1980-2005.* Rome: Food and Agriculture Organization of the United Nations.

Environmental code and legislative acts

The 2019 Forestry Code

The National Plan for Integrated Coastal Zone Management (PNGIZC)

New Environmental Code 2014

Law no. 96-766 of October 3, 1996

Law no. 2014-390 of June 20, 2014 on Energy Transition

Decree no. 2015-537 of July 16, 2015

Law no. 2015-532 of July 20, 2015 on the Maritime Code

Law no. 2019-700 of August 5, 2019 on the Protection and Management of Forests

Documents consulted for mangrove identification and classification

"African Mangroves

"Flora of Mozambique - Mangroves

"Mangrove Guidebook for Southeast Asia".

"Manual of Mangrove Species in Africa".

"World Atlas of Mangroves

"Mangroves of Western and Central Africa".

GLOSSARY

Mangroves: Tropical coastal ecosystems, mangroves develop in intertidal zones, offering diverse habitats and protecting coasts from erosion while contributing to carbon storage.

Rhizophora mangle (red mangrove): A mangrove species characterized by aerial stilt roots that stabilize the soil and reduce coastal erosion, it plays an essential role in saltwater filtration.

Rhizophora racemosa (white mangrove): Mangrove with stilt roots, this mangrove is resistant to salinity variations and forms dense forests in estuaries, helping to stabilize coastal soils.

Conocarpus erectus: Tropical wetland shrub, also known as "buttonwood", which grows behind mangroves and plays an important role in soil stabilization and coastal biodiversity.

Bruguiera: A tropical mangrove genus, Bruguiera species are renowned for their knee-shaped roots and their role in stabilizing the coastline and creating habitats for marine fauna.

Nephrolepis biserrata: A tropical fern common in wetlands, it thrives in mangrove undergrowth and contributes to biodiversity by serving as a habitat for many species.

Tidal zone: Coastal region subject to water fluctuations due to the tides, crucial for the dynamics of marine ecosystems and nutrient exchange.

Hydrology: Science that studies the distribution, circulation and properties of water on Earth, as well as its interactions with the environment and human activities.

Climate parameters: Factors such as temperature, humidity and precipitation that influence environmental conditions and ecosystems.

Water salinity: Measurement of the concentration of dissolved salts in water, a key parameter in determining the ability of aquatic ecosystems to support certain species.

Water temperature: An essential indicator of aquatic thermal conditions, it influences the biodiversity and metabolism of marine species.

Anthropogenic pressures: All human activities, such as urbanization and resource exploitation, which modify and often deteriorate natural ecosystems.

Printed by Books on Demand GmbH, Norderstedt / Germany